Modern Microscopy
Elementary Theory and Practice

Modern Microscopy
Elementary Theory
and
Practice

C.F.A. CULLING

Associate Professor, Department of Pathology, University of British Columbia, Vancouver, British Columbia; Lecturer in Microscopy and Histochemistry, University of British Columbia; Consultant Histochemist, Vancouver General Hospital, Vancouver, British Columbia. Formerly at Westminster Hospital School of Medicine, University of London; Lecturer in Histopathological Technique at the Sir John Cass College, University of London, and Bromley Technical Institute, Kent.

With a Foreword by
PROFESSOR W.L. DUNN, M.D., Ph.D.

Head, Department of Pathology
University of British Columbia;
Director of Laboratories,
Vancouver General Hospital

BUTTERWORTHS

ENGLAND: BUTTERWORTH & CO. (PUBLISHERS) LTD.
                    88 Kingsway
                    London WC2B 6AB

AUSTRALIA: BUTTERWORTH PTY. LTD.
                    586 Pacific Highway
                    Chatswood, NSW 2067

CANADA: BUTTERWORTH & CO. (CANADA) LTD.
                    2265 Midland Avenue
                    Scarborough, Ontario M1P 4S1

NEW ZEALAND: BUTTERWORTHS OF NEW ZEALAND LTD.
                    26-28 Waring Taylor Street
                    Wellington, 1

UNITED STATES: BUTTERWORTH (PUBLISHERS) INC.
                    19 Cummings Park
                    Woburn, MA 01801

©

ISBN 0 407 754008

Suggested UDC Number: 611—018·73:547·455

Suggested Additional Numbers: 611—018·25
                                      612·014·1

Contents

Foreword

During the past 25 years, with the expansion in commercial production of a wide-ranging variety of microscopes and the rapid development of techniques and applications, the obvious need for a compact but concise compendium of useful knowledge relevant to the art needs no documentation.

Through his long and detailed experience as a teacher in its theory and use, as an investigator employing the varied facets of its technical achievements, and as a consultant in many aspects of its more sophisticated use for over 20 years, Charles Culling is superbly suited to compile such a treatise on the microscope.

There are now hundreds of thousands of these instruments in their varied forms in use in the applied clinical, research and teaching fields throughout the world, employed in every branch of pure and applied science from basic chemistry and biology to industrial, fermentation, mining and the health sciences. New methodologies have emerged, the increasing complexities are legion, and the need for a text such as this combining a finely interwoven balance of basic theory in the use of lenses and optics with detailed instructions for their application and maintenance is timely.

I am not aware of any other single volume which in such succinct form considers lenses; their faults, corrections, types (ocular and objective), micrometry, dark field, phase and fluorescence (including fluorescent antibodies, their production, labelling and use) available today.

Detailed instructions and methodology are put forward in a lucid and concise style, dealing with comparison, binocular dissecting, polarizing, Nomarski phase and interference microscopes in addition to definitive chapters on the techniques and uses of photomicrography and the electron microscope.

FOREWORD

There is a very real need for instruction in the principles and methods involved in the varied aspects of microscopy based on personal knowledge of underlying theory, and applicable to a diversity of disciplines, both to professional investigator and student alike such as biology, chemistry and physics, pathology, engineering, as well as secondary school education and senior technologists.

I believe that the author's intent to make this text as widely useful as a handbook of microscopy has been achieved in its present form.

Department of Pathology,
University of British Columbia   Professor **W. L. DUNN, M.D., Ph.D.**

Preface

As a student of microscopy for the past 30 years, and a teacher of the subject for over 20 of them, I have been impressed by the two major points — the lack of knowledge of the subject by most of the users of the instrument, and the need for a complete, yet concise, practical textbook on the subject. I have endeavoured to write the latter in the hope it will diminish the former.

While I have made no attempt to deal completely with all the optical theory involved in explaining the various types of microscopes which are dealt with, I have included such elementary theory as is necessary to understand and use them intelligently.

The arrangement of the material in the various chapters has taken form over the years as my lectures have been modified to meet the needs of the various types of students, for example, medical laboratory technologists, biology and zoology students (both undergraduate and postgraduate), high school science teachers, and so on. It is aimed at providing the reader with complete and detailed instructions on how to select, use and care for the instruments described. The chapter on the Electron Microscope is the exception, since a chapter in such detail on this instrument alone would require a larger book than this is intended to be, and there are several excellent books already available.

I am indebted to Professor W. L. Dunn, Head of the Department of Pathology, University of British Columbia, for his interest, encouragement and constant stimulation.

I have been fortunate in being able to call upon my many colleagues for advice and criticisms. In particular, I would like to thank Doctor P. S. Vassar, and Doctor Philip E. Reid, who have been kind enough to read and criticize the script for me. I would particularly thank Doctor W. H. Chase who is mainly responsible for the chapter on Electron Microscopy, and Mr. B. J. Twaites who greatly assisted me with the chapter on Photomicrography.

PREFACE

These acknowledgements would not be complete without mention of my wife, Lois, without whose drive, criticism of the script and encouragement this book would never have been completed.

I would like to record my gratitude to Mrs. Audrey Spencer for secretarial assistance, to the many companies who kindly supplied illustrations, and to my publishers who have been a model of patience, tact and understanding at all times.

Vancouver, British Columbia   C. F. A. CULLING

Introduction

It is almost 700 years sincer Roger Bacon referred to convex lenses as an aid 'to old men and to those that have weak eyes'. Almost 350 years later Galileo (or Kepler, 1611) devised and used the first 'compound' microscope to examine insects. In 1673, Anthony von Leeuwenhoek, a Dutch draper and amateur microscopist, made the first of his many 'simple' microscopes with plano-convex and bi-convex lenses giving magnifications from X30 to X300; in 1674 he described his 'little animals' to the Royal Society of England, some of which are now thought to be bacteria.

Corrected lenses made their first recorded appearance in 1837, when Coddington produced his famous lens which was partially corrected. This was followed by the development of an achromatic substage condenser by Ernst Abbe in the 1870s, a type of condenser still in common use. Abbe who had been optical consultant to Zeiss since 1866 became a full partner in 1875 and sole owner, after the death of Carl Zeiss, in 1888 when he was aged 48 years. In 1932, Zernicke devised the principle of phase-contrast microscopy for which in 1953 he received the Nobel prize. In the previous year (1931) Knoll and Ruska had constructed the first electron microscope.

The interference microscope, the latest tool in microscopy, was devised and constructed by Dyson and Smith in 1952.

The compound microscope has already reached its theoretical limits of resolution, and it is only in the fields of utility and design that modern instruments show their improvement. Most post-war models (after 1946) focus by means of an adjustable stage and substage with a fixed body tube instead of the reverse; this has led to more pleasing designs and, because the stage is so much lighter than the body tube assembly gives a smoother operation with a longer life. The single control for both fine and coarse adjustment pioneered by Leitz has now been incorporated in a new form on the latest Reichart models. In the latter, the familiar large milled head still controls the coarse focusing, but a backward and forward movement of the whole spindle gives a smooth and easy fine adjustment.

The zoom type of lens, now so popular on cameras, was another post-war innovation, but it is mostly used today on low-power binocular dissecting microscopes where there is a greater need for minor variations in magnifying power.

With the increased use of phase and fluorescent microscopy, attachments have become available for their use on almost all makes and models. The number and variety of specially constructed condensers and prisms has increased so that mixed phase and fluorescence is now available from several manufacturers. This equipment enables one to use phase-contrast to identify a fluorescent cell, or tissue component, by the movement of a single control.

The automatic microscope which will take good photomicrographs in black and white (or colour) by simply pushing a button which was for so long the prerogative of Zeiss, is now available with, or as an attachment for, almost all the well known microscopes.

There are now available at a reasonable cost microphotometers for the quantitative determination of absorption and fluorescence intensity of microscope detail at various wavelengths; these are relatively simple in operation and give reproducible results.

Chapter 1

The Compound Microscope

The microscope is the most commonly used piece of apparatus in the laboratory, and yet it is probably the instrument about which least is known by its users. It is generally thought that the microscope can be used effectively without any knowledge of its limitations or construction, but this is, of course, a complete misconception. An ill-adjusted, badly illuminated microscope can, when one is using high-power objectives, give completely misleading information as to the structure of an object. For this reason it is advisable to gain a knowledge of how the magnified images are produced by the microscope before attempting to assess the information obtained by its use.

The first part of this chapter is devoted to the lens and its faults, after which the component parts of the microscope, its use and maintenance are discussed.

A LENS

A lens is the name given to a piece of glass or other transparent material, usually circular, having the two surfaces ground and polished in a specific form in order that rays of light passing through it shall either converge (collect together) or diverge (separate).

A lens is called positive when it causes light rays to concentrate or converge to form a real image *(Figure 1.1a–d);* or it is negative, in which case light rays passing through will diverge or scatter and positive or real images will not be seen *(Figure 1.1e–g)*. These two types are easily differentiated since positive lenses are thicker at the centre than at the periphery, whereas negative lenses are thinner at the centre and although the shapes may vary considerably, these characteristics remain *(Figure 1.1a–g)*.

Refraction of Light Rays

The effect of a lens on a ray of light is due primarily to the density of the glass (or other material) which reduces the speed at which light travels through it. (Light is usually considered as a vibration in the ether – a hypothetical substance which fills the whole of space). In a dense

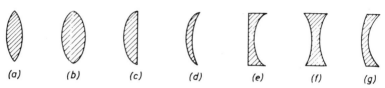

| (a) | (b) | (c) | (d) | (e) | (f) | (g) |

Figure 1.1 – Types of lenses. (a–d) Positive lenses; (e–g) negative lenses

medium (for example, glass) the light rays are retarded, or slowed down. If a beam of light containing two parallel rays (AB, $A_1 B_1$, *Figure 1.2a*) strikes a sheet of plane glass at right angles, its speed of travel through the glass will be reduced, but its direction unchanged. If it strikes the plane glass or the curved surface of a lens at an angle (CD, $C_3 D_3$, *Figure 1.2b*) its speed will be reduced and its direction changed. The bending of light rays, known as refraction, is due to the fact that one part (D) strikes the surface of the glass first (D_1) and is retarded while the other part (C) is still travelling at normal speed, thus causing the ray to be bent and its direction altered. From C_1 the two portions travel in the same direction at a common reduced speed until C emerges from the dense medium (C_2) and travels at its original speed while D is still retarded in the dense medium, causing a further bending of the light ray. After D emerges from the dense medium (D_2), the two portions travel in the new direction at their original speed.

It will be obvious that the degree of refraction will be dependent not only on the angle of the surface of the lens to the light ray, but also on the optical density of the material from which the lens is made. The optical density of a substance is indicated by its refractive index (R.I.), which is the ratio of the velocity of light in air to the velocity of light in that substance.

The behaviour of a beam of light passing from one medium to another can be estimated from the rule that light entering a more dense medium bends towards the 'normal' (A O in *Figure 1.2c*), and when entering a less dense medium it bends away from the normal (O B in *Figure 1.2c*).

A LENS

Light can always enter a lens, no matter what the angle at which it strikes, but it is not always possible for it to leave. As the angle between the beam of light leaving and the 'normal' (the angle of incidence) increases the emerging beam is bent closer and closer to the surface of the glass, until it is parallel with the surface (C O D in *Figure 1.2c*). Any further increase in the angle of incidence will result in the beam being reflected from the surface instead of emerging – a condition known as total internal reflection (E O F in *Figure 1.2c*).

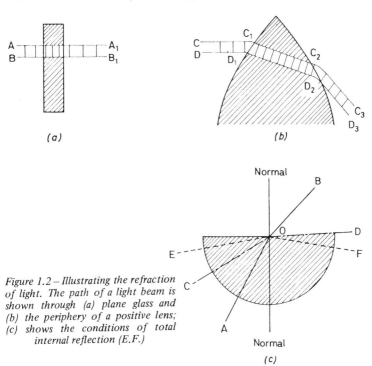

(a)

(b)

Figure 1.2 – Illustrating the refraction of light. The path of a light beam is shown through (a) plane glass and (b) the periphery of a positive lens; (c) shows the conditions of total internal reflection (E.F.)

(c)

Focus

If, through the centre of one side of a box, a pinhole is made, so small that only one ray of light can pass through it in each direction, then the image of an object outside the box will be formed on the back of the box *(Figure 1.3a)*. The ray of light from each point of the object entering the box is very narrow and it can only travel in a straight line. Therefore, each point of the object will have a corresponding point in the image *(Figure 1.3a)*, and since the light rays from the bottom of the object form the top of the image, and vice versa, the image will be inverted. Similarly, variations of brightness and colour will be reproduced. Such a box may be used as a camera, though not a very efficient one; a long exposure would be needed owing to the small amount of light allowed to enter. To enlarge the hole and fit a lens would result in the production of a much brighter image, owing to the fact that instead of only one light ray entering from each point of the object, a large number will enter through the fitted lens *(Figure 1.3b)*.

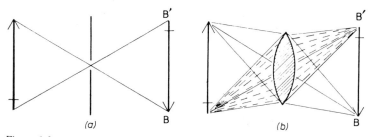

Figure 1.3 – Production of (a) image through pin-hole aperture, and (b) brighter image by using a lens instead of pin-hole

A notable difference in the production of images by a pinhole and a lens is that a pinhole will produce an image, regardless of the depth of the box or the nearness of the object, whereas in the case of a lens the

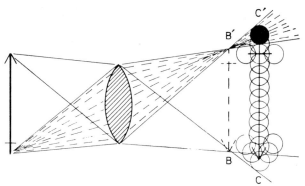

Figure 1.4 – The production of an in-focus image BB', and an out-of-focus image

screen and the object must be in exactly the correct positions, or the image will be indistinct and hazy. The lens will cause the light rays to converge to a single point at only one position (BB', *Figure 1.4*), and at either side of that position each point of the object will be represented by a solid circle of light; each circle being overlapped by the adjoining ones (CC', *Figure 1.4*).

When a lens concentrates the light rays to form a clear sharp image of an object, the object is said to be in focus. The terms 'focus' or 'principal focus' are used to indicate the position in which a lens will form a sharp, clear picture of a distant object, such as the sun. (The word focus originally meant burning place, and was used to indicate the point at which a lens concentrated the sun's rays to form a sharp image having the power to burn.)

In addition to the principal focus, a lens also has conjugate foci; these are two points, one on each side of the lens, in one of which a clear image will be formed on a screen of an object placed in the other. The positions of the conjugate foci vary: as an object is moved away from the lens, so the image will be formed closer to it and vice versa *(Figure 1.5a* and *b);* and any pair of such positions are called conjugate focal planes. The magnification of the lens is affected by this movement of the lens or object since the further away from the lens the image is

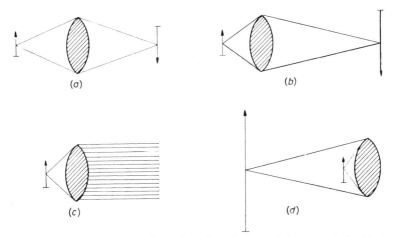

Figure 1.5. – *Showing the effect of moving a lens in relation to a static object (reproduced from the Microscope by courtesy of R. and J. Beck Ltd.)*

formed, the larger it appears *(Figure 1.5a* and *b)* and, consequently, the greater the magnification.

Images, as those described, which can be seen on a screen, are known as real images. As the object is brought closer to the lens, the image will move further away until it reaches infinity and cannot be seen *(Figure 1.5c).* If the object is brought still closer, the image will re-appear on the opposite side of the lens – that is, the same side as the object – but it will be a ghost image which can be seen only by looking through the lens, and which cannot be focused on a screen. The image has undergone a further change in that the image will appear the right way up *(Figure 1.5d)*. This is known as a virtual image.

Defects of a Lens

For a microscope to be efficient, it must not only produce a magnified image, but one which will be clear and well defined. To use a simple lens of the type described will not give such good results because: (1) white light is not a single vibration but is composed of a series of vibrations of differing wavelengths; and (2) faults are inherent in its shape.

Chromatic Aberration

When white light is split into its component parts, each part vibrates to a different degree, producing to the eye a different colour. These colours (red, orange, yellow, green, blue, indigo and violet) are known as the primary spectrum, and are seen in the rainbow, or through a spectroscope. Red has the longest wavelength, with a vibration of 0.7μ, blue 0.45μ and violet 0.35μ.

It will be seen that the vibrations of red light are twice the length of those of violet light.

Since light rays cannot vibrate as easily in a dense medium as in a rare medium, it follows that the various colours will be affected by a lens to differing degrees, the colours with shorter wavelengths, such as blue violet, being affected to a greater degree than those having a longer wavelength, such as red and orange. It is for this reason that rays of white light, having passed through a prism, emerge as a spectrum, each ray having been refracted to a different degree, and each emerging at a different point. Violet and blue are refracted to a greater degree than red and orange.

In *Figure 1.6,* W represents a point source of white light entering a lens which, on emerging, forms a different point of focus for each of the component colours, blue (B) being focused at a point nearer the lens than red (R).

A screen placed at R will show a red point surrounded by the colours of the spectrum, having a blue edge; at B a blue point with a red periphery will apear. This colour defect is called chromatic aberration, and its correction is known as achromatism.

Since different types of glass have different optical properties, chromatic aberration can be corrected to within useful limits by using a two-component lens. A positive lens (of greater magnifying power than is finally required) is combined with a negative lens made of glass producing a greater chromatic aberration, but with the same refractive index. The negative lens corrects the chromatic aberration in the positive lens, and only partially neutralizes its magnifying power. This method will correct a thin positive lens for any two colours, leaving a small error in the intermediate colours (secondary spectrum). This type of lens is known as an achromatic lens.

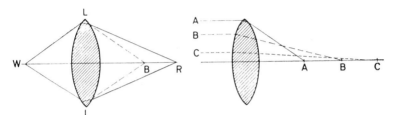

Figure 1.6 – Chromatic aberration   *Figure 1.7 – Spherical aberration*

If fluorspar is incorporated in the glass of the achromatic lens, three colours can be brought to the one focal point, and the amount of chromatic aberration visible in the image will be negligible. Such lenses are known as apochromatic lenses.

Since the correction of an apochromatic lens involves the use of a larger number of lenses, its other defects are corrected at the same time so that the final lens, provided that it is correctly used, will show hardly any defects.

Spherical Aberration

Spherical aberration is a further defect of a single lens, due to the fact that it has a curved surface.

Since the angle at which light rays enter (and leave) the surface of a lens varies with each part of the lens, those rays passing through the periphery (AA) will be refracted to a greater degree than those travelling through the central area (CC, *Figure 1.7*). There is no position, therefore, where the light from a point source will be in sharp focus, and since each point is hazy the composite image is bound to be indistinct. This fault could be minimized by using only the central area of a

7

lens, but since a microscope objective must have a short working distance and a high magnification, a large angle of light is required from each point of the object, and the correction of this aberration is most important.

The degree of spherical aberration will depend on the actual shape of the lens, and by varying the shape, although the focus may be the same, the spherical aberration will vary. The method of correction follows the same pattern used in correcting achromatism, namely, that of using a powerful positive lens and partially neutralizing its magnifying power with a negative lens made of glass having a greater relative aberration. Complete correction is extremely difficult and the above account presents the problem and its solution only in a very simple form.

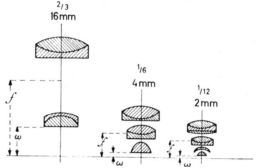

Figure 1.8 – Lens arrangement of the common objectives showing the relative focal lengths and working distances.
f = focal length; w = working distance
(By courtesy of R. and J. Beck Ltd.)

Chromatic and spherical aberration are the two principal faults to be found in lenses; there are others, but since correction is achieved by similar means – variation in shape and composition, and the distance apart of component lenses – they will not be discussed in detail. For a more comprehensive description of these faults the reader is referred to *The Microscope* by Beck (1938). Some idea of the complexity of various lens systems may be gathered from *Figure 1.8.*

COMPONENT PARTS OF A COMPOUND MICROSCOPE

A simple microscope is composed of one or several lenses mounted closely together, as in the case of a hand lens, whereas a compound microscope is composed of two widely separated lenses, or sets of lenses, capable of producing greatly enlarged images.

The standard monocular microscope *(Figure 1.9)* is composed of two main parts: (1) the microscope proper, incorporating the body tube with the objective at one end and the eyepiece at the other; and (2) the stand, which includes the supporting, adjusting and illuminating apparatus.

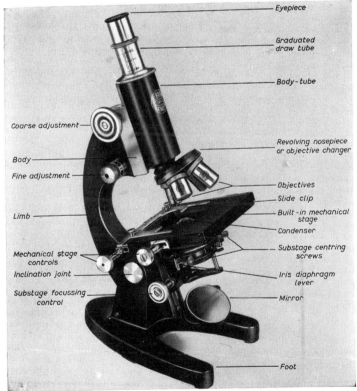

Figure 1.9 – The monocular microscope (By courtesy of C. Baker of Holborn Ltd.)

The Microscope Proper

The Eyepieces (or Oculars)

These are designed to further magnify the primary image from the objective, they also limit the field of view as seen by the eye. They may be used to correct residual errors in the objective lenses and may then be either: *under-corrected,* when a blue ray of light will be refracted to a greater degree than the red, this can be identified by the blue fringe that is seen around the edge of the field diaphragm; or *over-corrected,* when the reverse is the case and an orange fringe may be seen at the edge of the field diaphragm. Compensated eyepieces are usually over-corrected.

There are two basic types of eyepieces, as follows.

(1) With the *negative* eyepiece the focus is within (between) the lenses of the eyepiece. It is composed of two lenses; the lower or field lens collects the image that would have been formed by the objective (virtual image plane) and cones it down to a slightly smaller image at the level of the field stop (or field diaphragm) within the eyepiece

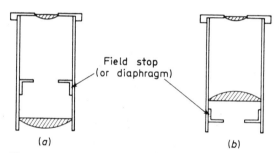

Field stop
(or diaphragm)

(a)   (b)

Figure 1.10 – Basic types of eyepieces (or oculars).
(a) Negative type (Huygenian), (b) Positive type (Ramsden)

(see Figures 1.10a and *1.16);* the upper lens then produces an enlarged virtual image which is seen by the microscopist. An engraved scale placed in the field stop will be superimposed (in focus) on the image (*see* the section on Micrometry on page 27).

(2) With the *positive* eyepiece the focus is outside the eyepiece lens system; for this reason it may be used as a simple microscope. The field stop (or diaphragm) is outside the eyepiece, from which the virtual image (from the objective) is focused and magnified by the entire eyepiece *(Figure 1.10b).* As with the negative type of eyepiece, a scale placed on the field stop will be superimposed (in focus) on the image formed by the objective.

Huygenian Eyepieces

These eyepieces *(Figure 1.10a),* originally designed by Huygens for the telescope, are the type most commonly used in microscopy. They are negative, under-corrected *(see above),* and are best suited for use with achromatic objectives.

Ramsden Eyepieces

As will be seen in *(Figure 1.10b)*, these are positive oculars. It will be noted that the lower lens has its plane side toward the object. Most of the compensated eyepieces are of the Ramsden type, having doublet or triplet lenses instead of the single lenses shown in *(Figure 1.10b)*. Ramsden oculars are preferred for micrometer eyepieces as they impart less distortion to scales.

Wide Field Eyepieces

Within recent years improvements in ocular design have enabled manufacturers to produce lenses which give a large flat field of view which are particularly valuable in the biological laboratory.

High-eyepoint Oculars

These have also been introduced in recent years, primarily for micro-scopists who wear spectacles, and are usually engraved with a diagram of a pair of spectacles. With normal eyepieces, the distance between the top of the eyepiece and the exit pupil (eye point) is so small as to prevent the wearing of glasses, but the high eyepoint of these special oculars make this possible. It is advised that the rubber guards supplied with such eyepieces be used to prevent the scratching of the spectacle lenses. Such eyepieces may be used by all microscopists, but some practice is needed before their use (with the head being held slightly higher than usual) becomes familiar and comfortable. The author uses ×12.5 high-eyepoint, wide-angle eyepieces routinely.

Compensating Eyepieces

These eyepieces were originally intended for use with apochromatic objectives only, and were not recommended for use with achromats. They are *now recommended for use with all modern objectives.* English speaking countries mark them 'Comp', while German lenses are desig-nated by the letter 'K'.

Field of View

Some eyepieces are marked with their field of view number from which can be calculated the actual diameter of the specimen being viewed (the field of view number, divided by the magnification of the objective, equals the field of view in millimetres). The figures for Zeiss eyepieces is given in the *Table* and may be used as a guide for other makes.

TABLE

Field of view and field of view numbers of Zeiss eyepieces

Eyepiece	Field of view numbers	Actual field in mm (with x10 objective)
x5	20	2·0 mm
x8	16	1.6 mm
x8 (K)	18	1·8 mm
x10	16	1·6 mm
x12·5	12·5	1·25 mm
x12.5 (wide field)	18	1·8 mm
x25	6·3	0·63 mm

Magnification

Eyepieces always receive the 'virtual image' from the objective in the same plane and therefore magnify it to a constant degree, independent of other factors such as body tube length, and so on. They are consequently marked with their magnifying power and may vary from X4 to X50. As will be seen in the following pages it is generally inadvisable to employ powers in excess of X12.5.

The Objective

The objective screws into the lower end of the body tube by means of a standard thread, thus all objectives are interchangeable. They are usually designated, not by their magnifying power but by their focal length (from 2 to 50 mm); this is because their actual magnifying power will depend on the tube length at which they are used. Some confusion has arisen in the past by the terms 'focal length' and 'working distance' in relation to objectives. Whereas with a simple lens these are identical, with compound lenses such as those in an objective they are different.

The 'working distance' is simply the distance from the object to the outer surface of the front lens, whereas the 'focal distance' is that from the object to a point roughly midway between the component lenses *(Figure 1.8)*. The latter is correct only when the objective is used at the standard tube length of 160 mm. If the tube length is altered the focal distance will also be altered and the object will need to be refocused.

Most instrument manufacturers mark objectives with the appropriate magnifying power, usually because they produce a microscope which has no draw tube, the tube length being a standard 160 mm.

The aperture. — The first objective consisted of a single lens, and its defects were overcome by the use of a pin-hole aperture, but since only a small cone of light could enter from each point of the object, the image, although greatly magnified, showed very little detail. It is apparent, therefore, that the amount of detail seen is dependent not, as commonly believed, on the magnification but on the size of the cone of light that can be collected from the object.

The ability of a lens to define detail is known as its resolving power, and this is measured by the distance apart of two lines or dots, or the number of lines to the inch, that can be visually separated from each other; for example, a lens that has a resolution of 30,000 lines to the inch has a greater resolving power than one separating only 20,000 lines to the inch. This will be appreciated by viewing *Figure 1.11a* from

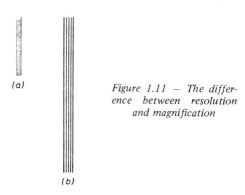

(a)

(b)

Figure 1.11 — The difference between resolution and magnification

a distance of 10 feet, when only a single line will be seen. On closer examination it will be found that there are, in fact, five lines. Even when the image is magnified *(Figure 1.11b)* and viewed from 10 feet it still has the appearance of being a single line

THE COMPOUND MICROSCOPE

Resolution is restricted by two factors: (1) the numerical aperture of the lens; and (2) the wavelength of light employed. The relationship is as follows.

$$\text{`x'} = \frac{1.2\,\lambda}{2\,\text{N.A.}}$$

Where 'x' is the resolution (the smallest distance between the closest two lines or dots that can be defined separately) and λ is the wavelength of the light employed. N.A. is the numerical aperture.

The numerical aperture. – The apertures of the early microscope lenses were at first measured by the actual angle of aperture; that is, the angle formed by the outer edges of the lens, and a point on the object *(Figure 1.12a)*. The aperture of oil-immersion lenses, however, depends

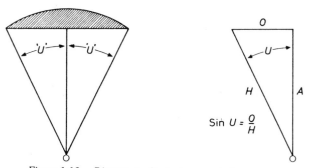

Figure 1.12 – Diagram to illustrate numerical aperture

on the refractive index of the medium between the object and the lens and for this reason may vary. This is because the cone of light, emerging from a glass coverslip into air, is refracted away from the lens face and a much smaller cone of light enters than if there was glass, or a medium having the same refractive index as glass, between the lens and the object. In *Figure 1.13a* it will be seen that the angle of the cone of light, from a point source of the object, actually entering the lens when used dry, is only 78 degrees, compared with an angle of 120 degrees when immersion oil is between them *(Figure 1.13b)*. To take account of this factor, and to be able to express a lens aperture as a simple figure, the term numerical aperture (N.A.) is used, which may be expressed as follows.

$$\text{N.A.} = n \sin u$$

where 'n' is the refractive index of the medium between the lens and object, and 'sin u' is the sine of half the angle of aperture *(Figure 1.12a)*. Since the sine of an angle is opposite over hypotenuse $\frac{O}{H}$ *(Figure 1.12b)*, it may also be roughly expressed as half the diameter of the lens over the distance from the periphery of the lens to the object. Since the highest N.A. theoretically possible when a lens is used dry (air R.I. = 1.0) must be 1.0, and with immersion oil (R.I. 1.51) 1.51, it will be appreciated that modern high-power objectives (dry N.A. = 0.95; oil immersion N.A. = 1.32) approach very closely to the theoretical maxima.

The effects of a high numerical aperture. — Whilst a high numerical aperture increases the resolution of an objective, it has the following disadvantages: (*a*) it reduces the depth of focus, that is, the ability to focus on more than one layer of an object at the same time; and (*b*) it

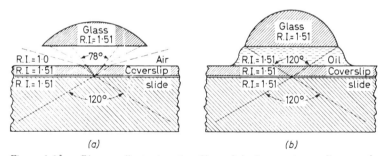

(a)   *(b)*

Figure 1.13 – Diagram illustrating the effect of the interposing medium on the angle of light entering the objective

reduces the flatness of the field, so that the edges are out of focus.

It will be seen, therefore, that if depth of focus and flatness of field are important, then increased magnification should be obtained with high-power eyepieces, although as a general rule a change to a higher power objective, giving increased magnification and resolution, is preferred.

Types of Objective

All objectives are engraved with the information needed to obtain their maximum performance as well as any possible limitations. Such an engraving might read: Plan 40/0.65

160/0.17

which indicates that it is a planachromat; X40 magnification at a tubelength of 160 mm (for its best performance), has a numerical aperture of 0.65 and should be used with a coverglass of 0.17 ± 0.01 mm in thickness (this /0.17 may instead be / = insensitive to coverglass thickness or /0 = for use with unmounted specimens).

Achromatic Objectives

These objectives are the type most commonly used and the modern well-corrected lenses are more than adequate for routine microscopy in pathology and biology laboratories.

Apochromatic Objectives

When apochromatic objectives are employed, their high degree of correction is wasted unless they are used in conjunction with a highly corrected aplanatic or achromatic condenser, and compensating eye-pieces. The latter compensates for peripheral chromatic aberration due to the differing magnifications of the various coloured images. These eyepieces together with the objectives form a re-combined single image free from coloured fringes even at the periphery.

Apochromatic objectives should always be used for microphotography. To get the maximum light with high-power objectives having numerical apertures above 1.0, oil-immersion condensers should be used with an N.A. at least equal to that of the objective, and immersion oil between the condenser and the slide as well as between the objective and the slide.

These objectives are also highly corrected for the other lens aberrations (spherical, coma, and so on).

Fluorite Objectives (Neofluor)

Fluorite or semi-apochromatic objectives, have fluorite incorporated into the lens system to give better colour correction. They are corrected for three wavelengths of light in the yellow-green of the spectrum, and are free of colour fringes. They are generally more highly corrected in all other respects than the achromats and represent a quality of image midway between that of the achromat and apochromat.

Planachromat Objectives

Planachromats are principally designed to give a perfectly flat field, with the whole field in focus at the same time. They are used mainly for photomicrography.

Polarizing Objectives

Designated POL, these are strain-free objectives for use on the polarizing microscope.

Phase Objectives

These objectives contain a phase-plate for use in phase-contrast microscopy (see page 93).

Coverglass Thickness

It will follow that oil-immersion objectives do not have coverglass restrictions since they will have the same refractive index as the immersion oil. The coverglass thickness is only important if high-power 'dry' objectives are being used, when No. 1 coverglasses should be used, or an objective with a correction collar may be employed which allows a range of thickness of coverslip from 0.12 to 0.22 mm to be used. To check the setting for a particular specimen (where the coverslip thickness is unknown) first focus upon a high contrast area, then determine whether changing the collar setting increases or decreases the contrast. If the coverglass thickness is known it can be set directly upon the engraved scale above the collar.

The Body Tube

The body tube is attached to the limb of the microscope which, in turn, is attached to the base either directly or by a hinged joint. Since the aberrations, or faults, of a lens can only be corrected for one tube length, for critical microscopy it should always be set to the standard 160 mm if a draw tube is fitted; if there is no draw tube, the body tube will, of course, be correct.

The body tube may rarely contain a draw tube, being a telescopic tube by means of which the distance between the eyepiece and objective may be varied *(see Figure 1.9)*. The draw tube usually contains a fixed diaphragm at its lower end to cut off reflections from the inside of the body tube. Such a draw tube is useful in micrometry (page 27).

A carrier or nosepiece for a number of objectives is usually fitted at the lower end of the body tube; it rotates on a central pillar, and is designated by the number of objectives it carries; for example, double, triple or quadruple nosepiece. The nosepiece should bring each objective into its correct position; that is to say, centred on the optical axis, and at the correct tube length. An increase in magnification is simply a matter of rotating the nosepiece, which is optically better than changing the eyepiece since a large aperture is being used; the oil-immersion lenses are, of course, an exception since the body tube needs to be raised to place oil on the slide.

The depth of the nosepiece will affect the tube length and this is generally 18 mm in depth, the actual length of the body tube being only 142 mm. If, for any reason, the nosepiece is removed, it must be replaced by a compensating ring of the same depth.

For accurate centring of objectives another type of objective changer may be used, a female slide being fitted to the bottom of the body tube, and each objective screwed into a male slide which has three centring screws. Owing to the improved design of modern nosepieces such attachments are now rarely seen.

Support, Adjustment and Illumination

Supporting Structure

The body tube of the microscope is attached to a limb, which in the past was hinged to a pillar and base. The latest models have inclined eyepieces, and the main supporting structure is not hinged.

Adjustment

On old models the body tube was attached to the supporting structure by two slides which were the site of the adjustment controls *(see Figure 1.9)*. This was followed by placing the slides (and controls) on the base, which entailed the controls moving the whole superstructure (body tube and limb) which caused increased wear and shorter life, but it was felt that the convenience of having the controls at almost bench level outweighed this disadvantage. Models of the past few years, however, have a fixed body tube, limb and base, the adjustment slide or slides being connected to raise and lower the stage and substage; this has the dual advantage, the controls being conveniently placed with little weight bearing on them, which gives longer life and lessens the likelihood of their 'slipping'.

The mechanism of the slides is such that one of them, working by rack and pinion, enables the stage and substage to be moved rapidly up and down, and is called the coarse adjustment; the other, working by micrometer screws, and levers or cams, enables the stage and substage to be moved slowly and accurately and is called the fine adjustment. Although the designs of the latter may vary, they are based on the same general principle: the movement by a lever or cam to a steel plate fixed on the back of the coarse adjustment slide *(Figure 1.14)*. The coarse

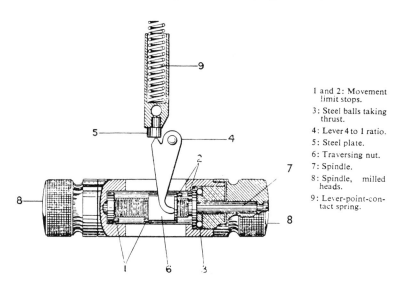

1 and 2: Movement limit stops.
3: Steel balls taking thrust.
4: Lever 4 to 1 ratio.
5: Steel plate.
6: Traversing nut.
7: Spindle.
8: Spindle, milled heads.
9: Lever-point-contact spring.

Figure 1.14 – A type of fine adjustment (By courtesy of C. Baker of Holborn Ltd.)

adjustment therefore moves the stage and substage, but the fine adjustment moves both the stage and substage and the coarse adjustment slide. As these slides wear, a degree of play will develop and cause slackness in focusing; most manufacturers, therefore, fit screws which may be adjusted to compensate for this slackness, but they should be adjusted with care as overtightening will cause excessive wear.

Stage

At the lower end of the limb supporting the body tube and adjustments is a platform, or stage, on which the objects to be examined are placed. The stage, which has an aperture of 1–1½ inches in diameter is provided with either simple metal spring clips to hold the object, which is then moved by hand to change the field, or is fitted with a mechanical stage which will give even steady movement of the object in two directions by means of two micrometer threads. The standard type of mechanical stage takes a 3 × 1 inch slide and moves over an area approximately 3½ × 1¼ inches so that a whole slide may be examined, but special stages are available to take very large slides and Petri dishes. Circular rotating stages are also available if preferred (for example, the polarizing microscope).

Most mechanical stages are fitted with a Vernier scale for recording the position of the slide in each direction, and they may be very useful if a particular field is to be found quickly at a later re-examination. By noting the reading on each scale, the slide can be replaced in much the same position almost immediately. One scale will be graduated, for example, from 0 to 80, and the other from 80 to 110 in order that the two readings will not be confused. Opposite these graduations will be the smaller Vernier scale, marked from 0 to 10. These 10 graduations, being equal to 9 in the main scale, enable each of the latter to be subdivided by 10.

Illuminating Apparatus

Below the stage, and usually attached to it, is an adjustable substage which can be moved up and down by a helical screw or rack and pinion (as is the coarse adjustment).

The substage consists of: (a) the condenser to focus the light on the object when using objectives with a focal length of 16 mm of less; (b) an iris diaphragm to control the cone of light entering the condenser; (c) a filter carrier; and (d) a mirror, flat on one side and concave on the other, which is mounted in gimbals so that light may be directed into the condenser from almost any angle, or more commonly a built-in light source.

The Condenser

The condenser should form a perfect image of the light source, and have the same numerical aperture as the objective with which it is being used.

The two-lens Abbé condenser is in common use but is not very efficient, forming only an imperfect image of the light source. It should not be used with apochromatic or fluorite lenses (page 16). To obtain perfect results with such objectives, a condenser with a lens system equal to that of the objective being used should be employed: a three-lens aplanatic or a more highly corrected achromatic condenser will give a crisp image with good resolution. Such condensers are usually fitted with a swing-out front lens (or the front lens may be unscrewed) to illuminate the whole field for low-power lenses. By swinging out the front lens the numerical aperture of the condenser is reduced to 0.3–0.4. For critical microscopy with objectives having an N.A. exceeding 1.0, immersion oil should be applied between the condenser and the slide, as well as between the objective and the slide.

The Iris Diaphragm

Light which passes through the object but does not enter the objective is unnecessary, and may interfere with those light rays which are intended to form the image.

The iris diaphragm is employed to limit the angle of the cone of light passing through the object so that it will just fill the front lens of the objective.

The intensity of illumination should always, if possible, be reduced by using light-absorbing filters, or a variable resistance, not by closing the diaphragm and never by racking down the condenser.

The Filter Carrier

The filter carrier is usually a recessed metal ring, pivoting on a screw to facilitate the easy removal of filters.

The Mirror

The two-sided mirror is plane on one side and concave on the other, and is fitted about 4 inches below the stage. A concave mirror has a focus since it causes the light rays, which have been reflected, to converge together and form an image. The focus is approximately 4 inches (its distance from the object) and is intended to take the place of a

21

condenser when using very low-power objectives since these require a large area of the object to be illuminated.

The plane mirror must always be used with the condenser since the latter can only be used efficiently if the whole of the back lens is filled with light.

The Binocular Microscope

The light rays emerging from the objective in the binocular microscope are equally divided between the two eyepieces. It is not sufficient simply to insert a single prism and divert one half of the rays, since this would cause eyestrain due to both the observer's eyes being focused on a single point a short distance away, and the advantage of a binocular

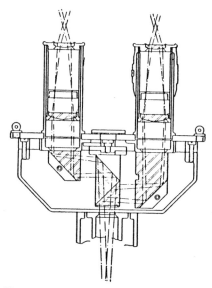

Figure 1.15 – The prism arrangement in a binocular attachment (By courtesy of Cooke Troughton and Simms Ltd.)

microscope is that long periods may be spent viewing through it with the minimum amount of eye fatigue. The modern binocular microscope achieves this by the use of four prisms. It will be seen from *Figure 1.15* that the eyes are receiving two parallel beams of light. The lower central prism consists of two prisms cemented together, at the interface of which there is a semi-silvered surface: this silvering is a very special process, fine grains of silver being deposited so that alternate light rays are differentially treated, one being reflected to the right *(Figure 1.15)* and the other passing into the upper prism.

The light rays passing through the semi-silvered surface to the upper prism travel through a greater thickness of glass than those that are reflected – having the effect of retarding them – and this is compensated for by making the right-hand prism with an extra thickness of glass as will be seen by comparing the two outside prisms in *Figure 1.15.*

An additional advantage of this system is that the eyepieces, with the prisms attached, can be easily moved together or apart, and the interocular distance adjusted to suit individual requirements.

One of the eyepieces (the one on the right in *Figure 1.15* is fitted with an adjustment to compensate for the slight variation of focus occasionally required.

With a binocular body on a microscope, the optical tube length may be increased from 160 to 240 mm, and since the objectives are corrected for the shorter tube length, a compensating lens is incorporated to overcome this factor; the lens is also necessary to re-focus the virtual image for the new tube length. The increase of tube length also has the effect of increasing the magnification, and binocular attachments may have their magnifying factor engraved on them which, since the tube length is usually increased by one half, is $\times$ 1.5.

Illumination

Although daylight may be used to illuminate the field, it will generally be found inconvenient owing to its inconstancy. Artificial illumination supplied by an electric filament lamp is therefore most commonly employed.

The lamp may be either a simple pearl bulb, or a high intensity lamp used in conjuction with a condenser and an iris diaphragm.

The source of illumination should be: (1) uniformly intense; (2) should completely flood the back lens of the condenser with light when the lamp iris diaphragm is open; and (3) make the object appear as though it were self-luminous.

(1) Uniform intensity of illumination is most difficult to obtain since the solid sources of light – a tungsten arc (where a small sphere of tungsten glows white), or a carbon arc – present great difficulties if used over long periods. The difficulty is overcome by using a closely wound filament with a diffusing screen, although for routine work with

a monocular microscope a 60 watt pearl bulb will suffice. Kohler illumination may be used.

(2)   The source of light should be sufficient to enable its rays when directed by the plane side of the mirror to flood the back lens of the condenser uniformly. The high intensity type of lamp has an optical axis and must be correctly aligned for use, and the distance from the microscope at which it is used adjusted so that the lens magnifies the lamp image to the correct size, a built-in light source has been so adjusted.

Where separate, the lamp and the microscope should be connected so that accidental movement of one or the other will not upset the alignment. If the manufacturers do not supply such a connexion, the lamp, the microscope and the transformer (if needed) may be mounted on a wooden base.

(3)   The object will behave as if self-luminous if the opal bulb or the image of the lamp condenser is focused in the object plane with the sub-stage condenser *(see below)*.

SETTING UP THE MICROSCOPE

The bench on which the microscope is mounted should be free from vibration and be in such a position that the microscopist works with his back to the window; a light screen, the back and sides of which are finished with a flat black paint to minimize back-scatter of light, is a great advantage.

Critical Illumination by Nelson or Kohler Methods

These are the two universally recognized methods for correct illumination.

Nelson Method

For this method the light source should be homogeneous and no lamp condensers used. It is normally employed with a bare light source.

Kohler Method

For this method to be used the light source does not have to be homogeneous, but a lamp condenser is essential to project an image of the lamp filament on to the substage iris diaphragm. In this system the lamp condensing lens (which is evenly illuminated) functions as the light source. This method must be used with compound lamps, and should always be used for micro-photography.

Technique

(1) The lamp should be positioned opposite the microscope (the high intensity compound light being fixed), and a blue daylight filter inserted in the filter carrier to absorb the excess yellow given by artificial light.

(2) Position the lamp so that the light strikes the centre of the mirror, and adjust the mirror so that the light is directed upwards into the condenser.

(3) With a compound lamp focus the condensing lens so that an image of the source of light is formed on the substage iris diaphragm; if necessary hold a piece of white paper at this position so that the image is visible. The daylight should be removed to get a clear image.

(4) Focus on an object on the stage with the 16 mm objective, and, with the eyepiece removed, adjust the mirror so that the field is evenly illuminated.

(5) Replace the eyepiece and, with the object in focus, rack the substage condenser up or down until a sharp image of the lamp iris appears; this renders the object self-luminous. In practice it has been found that the best position for the condenser is just below this point. This is 'Kohler illumination'.

(6) If the instructions given in (5) are followed with the exception that the condenser is racked up and down until a mark on the bulb, or the image of the lamp iris diaphragm, is focused, this will be *critical illumination* (Nelson method).

(7) Adjust the substage iris diaphragm so that only the area to be examined is illuminated; with objectives having a N.A. in excess of 1.0 the condenser diaphragm will need to be fully open.

(8) If the microscope is fitted with a centring substage, move the condenser up until a sharp image of the closed diaphragm is seen; the condenser is then adjusted until this image is central. Stages 4, 5, 6 and 7 are then carried out.

The microscope is now critically illuminated and the optical train co-axial.

For critical microscopy and microphotography, the lamp iris and the condenser may need to be re-centred each time the objective is changed.

One cardinal rule for the microscopist is always to rack the objective down near the object before looking through the eyepiece and then to focus on the object by racking the objective up and away from the object. This will avoid damaging the object, or the front lens of the objective, and is particularly important when using oil-immersion lenses, which have such short working distances.

THE COMPOUND MICROSCOPE

CLEANING AND MAINTENANCE

It must be remembered that the microscope is an exceedingly complicated and delicate piece of apparatus, and a great deal of experience
is required to completely service and maintain it. Component parts
should be returned to the manufacturer when faulty, since amateur
attempts at repair usually result in further damage: apart from cleaning
the outer surface of their lenses objectives are best left alone. Prisms
should never be touched, and cleaning should be confined to blowing
off the dust with a rubber bulb fitted with a small-bore metal tube,
since the slightest disalignment of the prisms will cause enormous eye
fatigue. Lenses should be wiped only with fresh lens paper or well-
washed silk, otherwise they may be scratched. Immersion oil should be
removed immediately after use, although old oil can be removed with
lens paper damped with xylol.

Daily Cleaning Routine

(1) The microscope should be dusted daily, and the outer surface of the lenses of objectives polished with lens paper or well-
washed silk.

(2) The top lens of the eyepiece should be polished to remove
dust or fingermarks, and the microscope set up for critical illumination.

(3) Rotation of the eyepiece will show if any dust is still present, in which case, the eyepiece may need to be dismantled and
both lenses cleaned.

(4) The substage condenser and the mirror are cleaned in a
similar manner: dust on the condenser will be apparent when this is
racked up and down, since it will come in and out of focus.

A little attention to cleaning the microscope daily will, by the removal of chemically-active and sharp pieces of grit and foreign matter,
prolong the life of the instrument and make the weekly cleaning task a
short and simple one.

Weekly Cleaning Routine

(1) The slides of the coarse adjustment, the mechanical stage
and the substage condenser should be wiped with a cloth damped
with xylol to remove dust which would otherwise damage the slides.
A little oil (as supplied for lubricating microscopes) is applied and the
slides replaced: later models do not require this treatment.

(2)   The lens system should be checked and cleaned.
(3)   Clean the eyepieces as described in (2) and (3) of the daily routine, and then trace dirt in other places by a similar system.
(4)   Dust is removed from the back lenses of objectives by use of the rubber bulb described above.
(5)   Interocular adjustment slides will usually require cleaning only once a month, and great care should be taken not to damage or disturb the prisms during this operation.

MAGNIFICATION

The magnification of a lens will depend on its conjugate foci (page 3); that is, the distance from the object to the lens and the distance from the lens to the image. In the microscope the objective forms a real inverted image in the upper part of the body tube, which is then further magnified by the eyepiece. Therefore, the magnification of the microscope is the product of the magnifications of the objective and the eyepiece, and is dependent on the following three factors: (1) the focal length of the objective; (2) the distance between the focal plane of the objective and the image it produces (since the optical tube length and the mechanical tube length are approximately the same, the latter is always used *(Figure 1.16)*); (3) the magnification of the eyepiece.
Magnification therefore equals:

$$\frac{\text{Tube length}}{\text{Focal length of objective}} \times \text{Eyepiece magnification}$$

To take an example: the magnification obtained with a 16 mm objective, used with a X 10 eyepiece at the standard tube length of 160 mm would be:

$$\frac{160 \text{ mm}}{16} \times 10 = 100$$

Where the magnification is marked on an objective it is only correct when used at the standard tube length. It should be remembered that this magnification is a linear one, and in the example above the object will be magnified 100 times in all directions; the actual area magnification will be 100 X 100 = 10,000 times.

MICROMETRY

The standard unit of measurement in microscopy is a micron (μ), which is 0.001 mm.
To measure microscopic objects an eyepiece micrometer scale is used

27

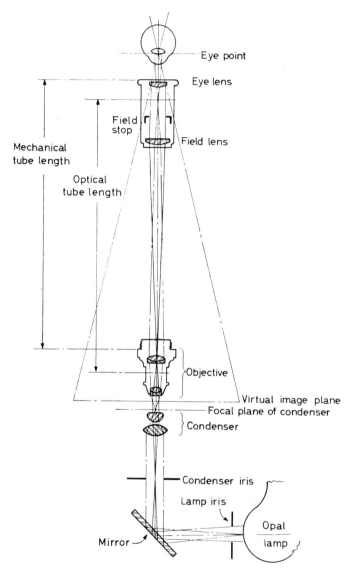

Figure 1.16 — Showing the optical path of light through a microscope

in conjunction with a stage micrometer. The eyepiece micrometer scale is usually a disc on which is engraved a scale divided into $^1/_{10}$ and $^1/_{100}$ graduations. This is placed inside the Huygenian eyepiece, resting on the field stop *(Figure 1.16)*. Eyepiece micrometers may be purchased with the scale permanently in position; these are usually Kellner eyepieces which have a focal plane below their bottom lens. They give a sharp image of the scale and have a greater eye clearance; they are an advantage (without a scale) for general work if spectacles are worn.

The stage micrometer consists of a 3 X 1 inch slide on which a millimetre scale is engraved in $^1/_{10}$ and $^1/_{100}$ graduations.

An object may be measured by the following method.

(1) Insert a micrometer eyepiece scale and place the stage micrometer on the stage.

(2) Select the objective to be used when measuring the object, and focus on the stage micrometer scale.

(3) Determine the number of divisions of the eyepiece scale equal to an exact number of divisions of the stage micrometer scale. A drawtube is useful at this stage since a slight alteration in magnification by increasing or decreasing the size of the stage micrometer scale, may greatly simplify calculations.

(4) Remove the stage micrometer, focus on the object to be measured, and determine the number of eyepiece divisions exactly covered by the object.

Calculate the size of the object as follows, assuming that 100 eyepiece divisions were equal to 10 stage divisions, and that the diameter of the object was exactly covered by 12 eyepiece divisions.

100 stage divisions = 1 mm = 1,000 μ
∴ 10 stage divisions = 100 eyepiece divisions = 100 μ
∴ 1 eyepiece division = 1 μ
∴ 12 eyepiece divisions = 12 μ

The diameter of the object, therefore, was 12 μ.

Chapter 2

The Comparison
Microscope

While the use of a comparison microscope for comparing bullets in police laboratories is well known, its use is comparatively rare in biological and pathological laboratories. Its use in those laboratories where the identification of species of plants, insects, ova, seeds, and so on, or of tissue entities by histochemical techniques is a daily problem, has usually not been considered. Yet placing reliance upon one's visual memory for even the time if takes to change slides upon a microscope stage is a dangerous practice, and the shortest route to a realization of this fact is to make several such comparisons and then examine the pairs of objects under a comparison microscope. The author currently (and for the past several years) spends some 10 hours a week examining histochemical preparations, which have been treated by different staining methods or chemical blocking techniques, with a comparison microscope *(Figure 2.1)*. By this method we are able to observe differences which would be almost impossible to see using conventional microscopy.

The Comparison Bridge

As will be seen in *Figure 2.1* the apparatus consists of two microscope bodies joined by an optical bridge. This bridge contains an arrangement of prisms whereby half the field of view from each microscope are brought together to form a bisected field, with the left half being from the left microscope and vice versa. While the simplest form of this type of microscope only performs this function, the model illustrated *(Figure 2.1)* is fitted with controls which permit either a half or full field from each microscope to be viewed simultaneously. Such controls allow the images to be superimposed for further comparison.

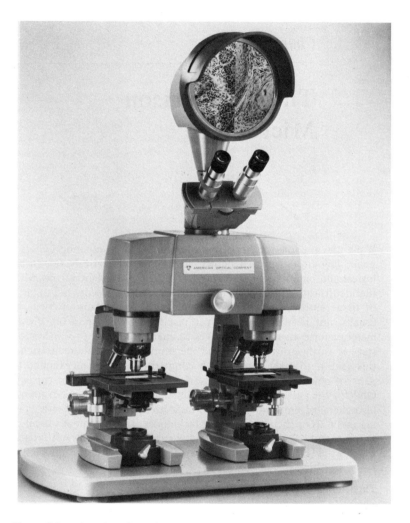

Figure 2.1 – American Optical Company Comparison Microscope with attached viewing screen.

Microscopes

While (if the bridge will fit) any two microscopes with similar optical equipment may be used; this arrangement will be found to be unsatisfactory. Identical models of the same make should be used, preferably with paired objectives and condensers. If the type of bridge used just slips into the drawtubes, then identical, paired eyepieces must be used. It will be found most difficult to achieve an even balance of light between the fields and, as will be appreciated, this is most important when comparing histological preparations to determine differences in staining reactions, which is quite often essential in assessing the effect of chemical blocking techniques. It is essential that identical light sources are used, each fitted with a variable control; a single light source used with a prism system to distribute the illumination evenly would be the ideal arrangement. The condensers should be properly focused and the microscopes critically illuminated (*see* page 24).

Mechanical stages are essential, and one or both should be rotatable in order that the specimens may be accurately aligned.

Transmitted light may be used with the reservations dealt with above.

Specimen holders for the comparison of solid objects which enables them to be rotated and/or tilted may be necessary for certain specimens.

Bullet comparison microscopes are available for this special purpose; for example, from Bausch and Lomb and from Leitz.

Chapter 3

The Low-Power Binocular Dissecting Microscope

The forerunner of the low-power binocular dissecting microscope was designed by H. S. Greenhough toward the end of the nineteenth century. It was a combination of two separate microscopes, mounted at an angle of 15–16 degrees to each other, fitted with matched objectives and focused by a single coarse adjustment. This type of microscope has certain inherent faults due to its design: (1) *it exaggerates the depth of an object or cavity;* and (2) since the optical axes of the microscopes are at an angle to each other their object planes are not parallel *and only the centre of each of the fields will be in sharp focus.* The latter is mostly compensated for by the great depth of focus and the fact that the focusing error is opposite in nature for each of the fields of view which results in a composite stereoscopic image appearing to be in sharp focus across the whole field.

Changes in design have resulted in the following.

(1) *Improved image quality* which reflects the advances in lens design.

(2) *Increased brilliance of image,* due to use of larger objective lenses.

(3) *Reversal of the image* by the use of prisms which, since it is now upright, makes dissection and orientation of the object more easy.

(4) *Increased range of magnification* by the introduction of rotary or sliding nose-pieces on which are mounted two, three or more paired objectives.

(5) *Further increase in the range of magnification* by the introduction of auxiliary lenses in the body tubes (for example, ×½, ×2) or by zoom attachments which give an infinitely variable range for each pair of objectives, or which may avoid the use of more than one objective pair.

THE LOW-POWER BINOCULAR DISSECTING MICROSCOPE

In practice this type of microscope offers a range of magnifications from approximately ×4 to ×100. Since the numerical aperture, due to the design, is limited to approximately 0.1, a magnification in excess of ×100 is usually 'empty magnification'. Their working distance is usually from 2 to 4 inches, although there are available some specially designed models with an extended working distance. They usually, but not invariably, have a built-in illuminating system which may be adjusted for incident or transmitted illumination of the object.

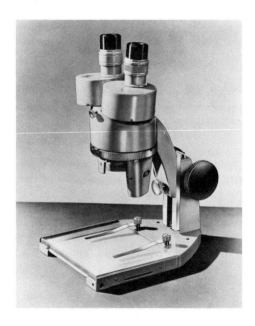

Figure 3.1 – Spencer stereoscopic microscope 26F (reproduced by courtesy of the American Optical Corporation)

These models should allow easy adjustment of the interpupillary distance and be fitted with an adjustable focusing collar on one of the eyepiece tubes to allow for inequalities in the observer's vision. A variety of stages (glass, reversible black and white, recessed, and so on) are available and it (or they) should be chosen with a view to the material likely to be examined.

The models shown in *Figure 3.1* and *3.2* illustrate features that are available from other microscope manufacturers. Most of the available models will have certain features specific to that model (or manufacturer) and before purchasing such an instrument it is wise to prepare a list of the features required for the investigations to be performed. An instance of the problems to be encountered are that with a ×20 magnification there will be available working distances from 1.5 to 8 inches,

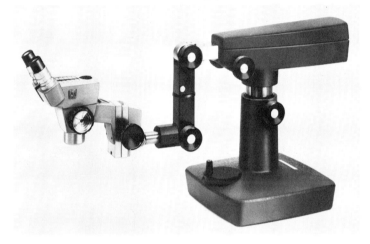

Figure 3.2 – Cycloptic microscope on universal stand (reproduced by courtesy of the American Optical Corporation)

with a field of view varying from 0.3 to 0.5 inches with varying numerical apertures. A decision must be made as to the relative importance of resolution, working distance, magnification and field of view with their importance being dependent upon the use to which the microscope will be put.

Chapter 4

The Dark – Ground
Microscope

For an object to be examined microscopically, it must first be visible. Visibility is dependent on contrast, as is illustrated by the fact that a black cat is invisible in a coal cellar because there is no contrast between the object and the background. For the same reason, a spider's web is difficult to see against the sky, yet stands out clearly when viewed against a dark background with the sun shining on it; this is because the fibres reflect the rays of light from the sun and give the web the appearance of being self-luminous, the dark background increasing the contrast.

Most objects examined microscopically are naturally transparent, but in general they reflect or scatter light rays, and if, as in dark-ground illumination, oblique light is thrown upon them which does not enter the objective, they will appear as self-luminous objects on a dark background.

Objects examined by dark-ground illumination give a misleading impression of size; fine particles appear to be much larger than they are, owing to their light-scattering properties. This factor is of advantage when examining fine structures such as spirochaetes which are clearly visible by this method, yet when stained (by Giemsa's stain) are difficult to see. This will only apply if the object is alone or nearly alone in the field of view; therefore, preparations must be as thin as possible; if such objects are examined in a mass of light-reflecting material the contrast will be lost. Although it is impossible to completely isolate cells and organisms, extraneous refractile material such as air-bubbles, red blood cells and oil droplets must be avoided and a thin preparation used.

Objectives and Condensers

Low-power objectives work at some distance from the object and therefore dark-ground illumination is obtained simply by inserting a small circle of black paper (pasted on glass) in the filter carrier. The central rays which would normally pass through the object and into the objective are cut off and the peripheral rays from the condenser pass through the object, but do not enter the objective; the only light entering the objective will be that scattered by the object.

High-power objectives, having a much shorter working distance require a special condenser which will accurately focus a hollow cone of light at an acute angle. This angle is so acute that if oil is not used between the condenser and slide the light rays are reflected back into the condenser (total internal reflection, *see* page 3). Immersion oil must be used between object and objective to ensure that the maximum

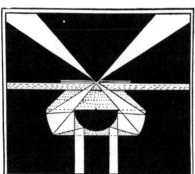
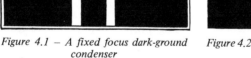
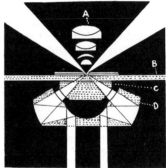

Figure 4.1 – A fixed focus dark-ground condenser

Figure 4.2 – A focusing dark-ground condenser

(Reproduced from 'The Microscope' by courtesy of R. and J. Beck, Ltd.)

amount of reflected light from the object enters the objective. To get the best results the condenser must be accurately centred, otherwise peripheral light rays will enter one side of the objective; similarly, the condenser must be accurately focused to get the maximum amount of light on the object without its entering the objective.

Because of the very acute angle of light required, very few dark-ground condensers can be used with an objective having a numerical aperture (N.A.) in excess of 1.0. A 2 mm objective having a N.A. 1.3 can be used if a funnel stop (a small metal tube) is inserted in the back which reduces the working aperture to less than 1.0. Alternatively, a $1/7$-inch (X70) oil immersion lens may be used without modification. The most convenient type of 2 mm objective is one incorporating an iris diaphragm, since this can be closed just sufficiently to stop any direct light.

The fixed-focus type of dark-ground condenser *(Figure 4.1)* is most common, but this can only be used with extra thin glass slides and coverslips (No. 1). Focusing dark-ground condensers *(Figure 4.2)* are available which will allow a variety of slides and coverslips to be used.

Since only reflected or scattered light forms an image of the object, the source of light should be an intense one, to ensure the maximum amount of light passing through the object. A Pointolite tungsten arc lamp probably gives the best results, although the modern high intensity lamp will give almost equally good results.

SETTING UP THE DARK-GROUND MICROSCOPE

Method

(1) Make a thin preparation, using a thoroughly clean thin slide and coverslip, and taking care not to have air-bubbles in the preparation.

(2) Place the lamp in front of the microscope and focus the image of the source on the plane side of the mirror (if necessary).

(3) Direct, or adjust, the light through the condenser so that it is evenly distributed.

(4) Rack the condenser down; place a drop of immersion oil on the top lens of the condenser and on the lower side of the slide. Place the slide on the microscope stage and slowly rack up the condenser until the two surfaces of the immersion oil meet without forming air-bubbles; such bubbles would reflect light in all directions.

(5) Focus on the object with a low-power objective such as the 16 mm.

Note. — If the condenser is correctly focused a small point of light will illuminate the object on a dark background. If a hollow ring of light is seen the condenser is above or below its point of focus and should be adjusted.

(6) With the centring screws, adjust the condenser until the point of light is in the centre of the field.

(7) Place a drop of immersion oil on the coverslip and focus the object with the high-power oil immersion objective. Perfect dark-ground illumination should result if a funnel-stop objective is used; if an iris diaphragm is incorporated in the objective it is adjusted to give the maximum performance. Occasionally the objectives are not

par-central and the condenser may need a slight adjustment to get a perfect result.

(8)   After use the oil should be carefully cleaned off both the condenser and objective.

The following errors are the most common causes of difficulty in setting up the microscope.

(1)   The slides or coverslips are too thick.
(2)   The preparation has too many air-bubbles present.
(3)   Condenser is not properly focused or centred.
(4)   Lighting is not sufficiently intense.

Chapter 5

The Fluorescent
Microscope

In 1852 Stokes first used the word 'fluorescence' to describe the reaction of fluorspar to ultraviolet light: in 1903 R. W. Wood devised a filter which would absorb visible light and transmit only ultraviolet light. These two events led to the first 'fluorescence microscope' described by Lehmann in 1911. Little use was made of this apparatus until 1935 when Max Haitinger pioneered and developed the technique of staining histological preparations and smears with fluorescent dyes. It is probably to him that most of the credit for the modern development of fluorescence techniques belong. In 1937 Hageman applied fluorescent dyes to organisms, and probably the first routine use of fluorescent microscopy was the staining of acid fast bacilli. In 1941 Coons, Creech and Jones described a technique for labelling protein with a fluorescent dye, which led to the now almost routine technique of fluorescent antibody staining.

Fluorescence

When a quantum of light is absorbed by an atom or molecule, an electron is boosted to a higher energy level. When this displaced electron returns to its original ground state it may emit a quantum of light *(Figure 5.1)*. If this light is emitted only during the time of exposure, or for a very short time afterwards (about 9–10 seconds) it is known as *fluorescence*, if the emission persists after the exciting light is cut off it is called *phosphorescence*. Since a certain amount of energy is lost as heat before the electron returns to its ground state the fluorescent (or phosphorescent) light is at a longer wavelength (lower energy) than the original exciting light. In fluorescence microscopy ultraviolet light (which is not visible to the human eye) is used as the exciting light with

43

the resulting fluorescence (of a longer wavelength) being in the visible range. Thus an object is illuminated with 'black' light and, when fluorescent, appears as a bright object on a dark background. It should be remembered that while an enormous number of compounds are

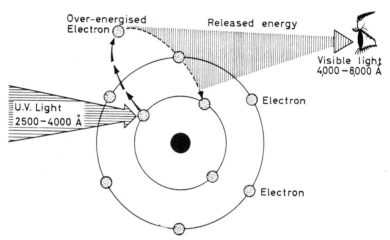

Figure 5.1 – Diagram to illustrate how ultraviolet light may excite fluorescence in a molecule

fluorescent to some degree, only relatively few give sufficiently brilliant fluorescence that they may be detected in small quantities by their autofluorescence (*see* 5HT page 56), or used as fluorescent dyes. Certain dyes, marked in catalogues as fluorescent, are virtually useless because of their poor fluorescence. Furthermore some compounds and dyes, while brilliantly fluorescent as pure compounds, may lose their power to fluoresce when bound to other structures. This is known as *quenching* of fluorescence. This latter is sometimes a useful property, since non-specific fluorescence can be quenched to give greater contrast (*see* use of haematoxylin to quench nuclear fluorescence, page 54).

Equipment

The early workers in this field used quartz condensers and slides to bring the maximum concentration of ultraviolet light on to the object. With modern optical glass (which transmits light with a wavelength of over 300 mμ) it has been found that almost any condenser may be used for this purpose.

Slides should be checked for obvious fluorescence (those made of green glass being avoided); but most good brands of slides are suitable.

The only special equipment that needs to be purchased, provided a microscope is available, is a good ultraviolet light source, special light filters and ideally, though not necessarily, a polished metal (surface reflecting) mirror.

Illumination

Any good ultraviolet light source may be used, for example, carbon arc lamp, Mazda Mercra lamp, and so on. The ones most commonly employed are dealt with in detail below.

Osram HBO 200 Lamp*

This is a high pressure mercury lamp which provides a steady powerful source of ultraviolet light. The mercury arc, measuring 2.5 by 1.3 mm, operates in a globe of fused quartz at a pressure of 70 atmospheres with a brightness of 25,000 stilbs (25,000 candle power per square centimetre). The starter unit provides 15,000 volts to strike the arc, which is maintained by low tension of about 60 volts. It should always be employed within the special housing provided.

This lamp is without doubt the most efficient and least troublesome of those available. With the exception of the AH6 *(see below)* it gives the most intense illumination. It gives five-sixths the light intensity of the AH6, but since it does not require water cooling it is much more convenient to use and there is no installation problem. Most microscope manufacturers (Zeiss, Leitz, Reichart) now incorporate it in their fluorescence equipment. The lamp has an average rated life of 200 hours, but the author used his first one for over 2,000 hours; his second lamp (in 4 years) is still in constant use. There is a fall in emission with wear but this was not sufficient to make any practical difference. The only rule followed was that once the lamp was switched on it was left on for the whole of that working day. The lamp gives over 30 per cent of its emission at a wavelength of 365 mμ which is the wavelength found to be most useful in fluorescence microscopy *(Figure 5.2)*.

*In the U.K. marketed under the name 'Neron'.

100 Watt, Type AH4 Mercury Vapour Arc
(General Electric Co.)

This lamp is a fused quartz tube, with sealed-in tungsten electrodes, giving an arc stream about 25 mm in length and 1.6 mm in width. The average rated life of this lamp is 1,000 hours.

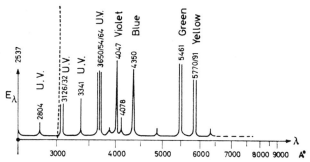

Figure 5.2 – Emission spectrum of the super pressure mercury lamp Osram HBO 200. The interrupted line at 3,000 Å indicates the transmission limit of optical glass; light of a shorter wavelength than this can only be used with a quartz condenser

The AH4, available from a number of microscope manufacturers, is employed in many laboratories for routine fluorescence work. It is less expensive than the HBO 200, and has a longer life. It is not suitable for critical fluorescence, for example, fluorescent antibody technique or autofluorescence assessment, but is adequate for most fluorescent staining methods.

1,000 Watt, Type AH6, Water Cooled Mercury Lamp
(General Electric Co.)

This lamp consists of a quartz capillary tube, about 2 inches by ¼ inch. The arc stream is about 25 mm by 1.6 mm. A special transformer supplies an operating voltage of about 840 volts. It develops tremendous internal pressure which necessitates its encasement in a water-jacket to develop a counteracting pressure; this jacket also functions as a cooling system. For efficient operation of the lamp a pressure operated switch and magnetic valve must be inserted into the water line which must be attached permanently to the lamp.

This lamp is without doubt the best available for fluorescence microscopy since it has the highest intrinsic brilliance of any artificial illuminant, with a continuous emission spectrum. However, the cost of installation and the loss of mobility in most instances outweigh the slight advantages.

Projection Lamps

These have been used by some workers, who utilize light in the blue–violet visible range (400–450 mμ) to excite fluorescence in the green–yellow (550–600 mμ) range. They are not recommended for general use except where no other source is available.

Filter System

A heat filter system is essential with any intense source of illumination. The heat filter is usually located in the lamp housing between the lamp and the collecting lens.

In addition, two basic filter systems are necessary for fluorescence microscopy:

(1) *Exciter filters.* – These transmit light of a short wavelength to excite fluorescence in the specimen. They may be varied to transmit (*a*) light of a specific wavelength, (*b*) light up to 400 mμ (ultraviolet), or (*c*) light up to 500 mμ (ultraviolet-blue).

(2) *Barrier, or contrast filters.* – These are so named because they are used primarily to protect the eyes from the damaging effects of ultraviolet light. By the use of different filters, with varying absorption and transmission characteristics, non-specific background fluorescence may be extinguished, giving greater contrast. For example, when examining tissue stained by a yellow or orange fluorescent dye, Schott filters OG 4 and 5 (Zeiss 47 and 50) may be used to absorb the blue autofluorescence of the tissue. This will result in bright yellow staining against a dark background.

Exciter Filters

For routine use a 2 mm Schott BG 12 filter (325–599 mμ) should be used in conjunction with a dark ground condenser. With a bright ground condenser a 4 mm thick BG 12 filter (or even 6 mm) should be used.

For special purposes a Schott UG 1 (275–400 mμ) may be used alone, or in combination with 2 mm BG 12 (325–400 mμ). This latter combination is recommended for use in fluorescent antibody staining techniques, although with minimal fluorescence the 2 mm BG 12 or UG 1 alone may be preferable.

Other filters *(Figure 5.3)* alone, or in combination, may be used for specific wavelengths.

Barrier Filters

These may be used to specifically absorb light below a given wavelength *(Figure 32.4)*. Zeiss filters are now numbered 41, 44, 47, 50, 53, to designate the wavelength at which they transmit light, for example,

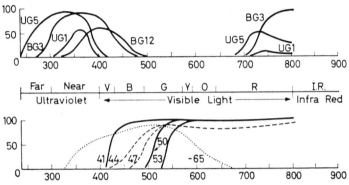

Figure 5.3 – Diagram showing transmittance curves of (top) exciter filters and (bottom) barrier filters. Figures on the ordinate scale indicate the percentage of transmission. Wavelength is shown in millimicrons (mμ)

41 will transmit light with a wavelength of 410 mμ and above. The filter − 65 is used to absorb unwanted red light (above 650 mμ) which is transmitted by several of the exciter filters.

Fluorescent staining techniques. − The most useful filters are the yellow Schott OG 4 and orange OG 5 (Zeiss 47 and 50) used alone or in combination.

Fluorescent antibody techniques. − A pale green–yellow GG 9, Zeiss 44, or Euphos filter is used.

Autofluorescence. − A colourless ultraviolet stopping filter is used simply as a barrier filter (Schott GG4, Zeiss 41).

Any make of filter may be used; they are ordered by specifying the transmission wavelength required.

Microscope

Any good microscope may be used for fluorescence microscopy. It is often convenient to purchase the lamp (HBO 200) from the manufacturer of the microscope being used since there will probably be convenient points of attachment to set the lamp in the correct position (distance).

Mirror. — This should be of a front surface reflecting type (polished metal) to avoid loss of ultraviolet by double surface reflection (for example, glass face and mirror face) and to avoid the possible absorption of ultraviolet, by the glass. However, it will be found that a large number of normal microscope mirrors will give satisfactory results.

Condenser. — A *light type condenser* may be used, particularly with low-power objectives; a *dark ground condenser,* however, is almost mandatory for oil immersion objectives, since it gives a darker background and allows a thinner exciter filter to be used. The disadvantage of a dark-ground condenser is that oil must be used between condenser and slide (*see* page 40) but this is found to be far outweighed by the advantages. In practice the author uses a dark ground condenser as a routine.

Contrast-fluorescence condenser. — This combined fluorescence–phase condenser is available from Reichert, and is intended for use on their Binolux microscope. The specially designed condenser annulus (the whole of which passes ultraviolet light) permits examination by phase contrast, fluorescence, or a mixture of the two. It is most useful as a means of identifying the source and location of fluorescence in smears and sections.

Barrier filter attachment. — Barrier filters may be as follows.

(1) Inserted into the eyepiece by removal of the top lens, or they may be screwed into the bottom of the eyepiece.
(2) Inserted in the body tube by means of specially fitted slides (carrying one, or a number of filters), or by placing a single filter in a convenient location.
(3) Incorporated in a rotary filter changer (such as that supplied by Zeiss) which is fitted below the binocular attachment.

If the microscope is used for a variety of purposes type (3) will be found the most convenient, since filters are easily and quickly changed.

Since two rotating discs each carry 3 filters and one blank space, one can use a variety of filters, either alone or in combination.

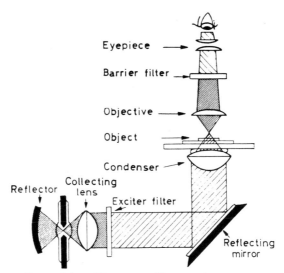

Figure 5.4 — Diagram to illustrate the component parts of the fluorescent microscope. The lamp gives out mixed ultraviolet and visible light (darkened area); the visible light is filtered out by the exciter filter. The object gives rise to visible light which is mixed with ultraviolet light (darkened area); the ultraviolet light is filtered out by the barrier filter, so that only visible light reaches the observer's eye

Objectives. — Any non-fluorescent objective may be used. Achromats are generally preferred to apochromats as they rarely fluoresce and their colour correction is usually adequate. A high numerical aperture (N.A.) is preferred to ensure the maximum transmission of fluorescent light from the object. The oil immersion objective should be fitted with an iris diaphragm (or funnel stop) when using a dark ground condenser.

Microscopic Preparations

Microscope slides. — These should be thin and of even thickness (not of green glass). Special ultraviolet transmitting slides may be purchased, but unless a quartz condenser is used it is pointless to employ them. Optical glass (as used in condensers) will only transmit light of 300 mμ and over, and at this range thin glass slides have an adequate transmission.

Section adhesives. – Thinly applied routine section adhesives do not interfere with preparations.

Mountants. – Cleared preparations may be mounted in H.S.R. (Harleco synthetic resin) or Depex. Fluormount will probably give the best results.

Aqueous mounts. – These may be mounted in Apathy's media with the exception of acridine orange or fluorescent antibody stained preparations.

Fluorescent antibody preparations. – These are mounted in glycerin to which 10 per cent phosphate buffered saline (pH 7.1) (*see* page 76) has been added.

Acridine orange stained preparations. – These are mounted in buffer only.

AUTOFLUORESCENCE (PRIMARY FLUORESCENCE)

The ability of some naturally occurring compounds to fluoresce is on occasion a great advantage in identification. Autofluorescent material can present a great hazard to the inexperienced microscopist, because, dependent on its structure, it may fluoresce any colour and thus appear to have been stained by the technique employed. For this reason unstained smears, identically prepared in all other respects, should always be used as controls of fluorescent stains.

Preparation of Material

For the specific study of autofluorescence, unfixed smears or cryostat cut sections of unfixed tissue should be used. It may be found subsequently that fixation does not interfere with the specific fluorescence. 95 per cent alcohol (ethyl) or ether–alcohol are usually satisfactory. Formalin should generally be avoided if possible as it tends to increase the blue autofluorescence of tissue, however with 5HT *(see below)* it is essential.

Specific Autofluorescence

The number of naturally occurring autofluorescent compounds is enormous and for a more complete list of them the reader is referred to *Fluorchemistry* by De Ment (1945) (Chemical Publishing Co.), or *Fluorescent Analysis in U.V. Light* by Radley and Grant (1951) (Chapman and Hall). Those dealt with below are considered the most likely to be encountered.

Tissue. Generally fluoresces a bright blue, although this may be absorbed by use of a yellow or orange filter.

Elastic fibres. – Fluoresces an intensely brilliant blue while unstained, and may be easily seen even in an H. and E. stained section.

Ceroid and riboflavine, – These fluoresce in shades of yellow.

Lipids and lipochromes. – Many of these fluoresce in shades of yellow.

Vitamins. – Many vitamins are fluorescent in shades of yellow, green and blue.

Porphyrin. – This group (and chlorophyll) are among the very few compounds with an intense red fluorescence (Hellstrom, 1934). This characteristic has been made use of by adding a drop of concentrated H_2SO_4 to blood stains (or suspected stains), the H_2SO_4 takes the iron out of the haemoglobin forming haematoporphyrin which gives a brilliant red fluorescence. There is a small accessory lacrymal gland (Harderian) in the corner of the eye of some animals which, having a high porphyrin content, gives this characteristic fluorescence.

Nissl substance. – Fluoresces a bright yellow colour in formalin-fixed unstained sections.

5-Hydroxytryptamine. – Gives golden yellow fluorescence (in argentaffin or enterochromaffin cells) after formalin treatment.

Drugs. – Certain drugs give a characteristic fluorescence. The ability of *tetracyclines* (terramycin) to form bright yellow fluorescent foci in malignant tumours has been investigated (Vassar, Saunders and Culling, 1960). The author has also used this antibiotic (since it is bound by calcium) to show areas of new bone formation in tetracycline fed animals.

Hydrocarbons. — The carcinogenic compounds, in particular, have been found to be strongly fluorescent. Vassar, Culling and Saunders (1960) utilized this method to demonstrate their presence in histiocytes in sputum from heavy smokers. 3:4 Benzpyrene has been used by Berg (1951) to demonstrate even the finest lipid granules (*see* page 55).

FLUORESCENT STAINING TECHNIQUES

The number of fluorescent staining techniques which are applicable in a routine laboratory are still moderately few. Those methods described below have been tried in the author's laboratory and were found to be reasonably reliable. A great number of fluorescent techniques recently described have great promise for the future, particularly those in enzyme histochemistry which will enable specific demonstration of minute areas of activity. These more recent techniques have yet to be evaluated for routine work and are not therefore included.

Fluorescent antibody techniques are dealt with on page 65.

Fluorescent Stain for Amyloid

This stain, which was developed following research on twenty-six fluorescent dyes under varying conditions (Vassar and Culling, 1959), has proved to be the most specific of those stains currently available. It has the additional advantage of not requiring microscopical differentiation. It has on many occasions demonstrated amyloid when all other methods have failed; in at least three of these cases the patients have subsequently died and the diagnosis was confirmed. Because of the sensitivity of fluorescence technique even the finest deposits of amyloid are seen. The only other tissue components that stain are mast cell granules and myeloma casts in the kidney (Vassar and Culling, 1962).

In practice it will be found that a typical amyloid structure is easily recognized. The technique is simple, the reagents stable and the method absolutely reliable. The stained slides may be examined using the BG 12 exciter filter with an OG 4 and/or OG 5 barrier filters which give a bright yellow on a black background; but the use of a UG 1 or UG 2 exciter filter with a colourless U.V. filter gives a brighter yellow on a blue background and will show the finest amyloid deposits (for example, heart).

Fixation is not critical. We use formalin-fixed, paraffin embedded material. Frozen or cryostat sections may be used.

Method

(1)   Bring sections to water.

(2)   Stain in alum haematoxylin for 2 minutes, to quench nuclear fluorescence. The haematoxylin does not need to be differentiated, or blued.

(3)   Wash in water for a few minutes.

(4)   Stain in 1 per cent aqueous thioflavine T for 3 minutes.

(5)   Rinse in water.

(6)   Differentiate in 1 per cent acetic acid for 20 minutes.

(7)   Wash in water.

(8)   Mount in Apathy's medium.

Results (*see above* for details of filters employed)

Amyloid	Bright yellow
Mast cell granules	Yellow

Acridine Orange Staining Techniques

The original method for differentiation of RNA and DNA is described on page 62. In this connection it should be remembered that proof of RNA and DNA structures must be demonstrated by the use of enzymes. Acridine orange has been shown to demonstrate mucopolysaccharides although in the author's experience this is rare except with the technique given below.

Hicks and Matthaei (1958) discovered that a section previously stained by iron haematoxylin would, if subsequently stained with acridine orange, demonstrate mucins with some degree of specificity. The author made a similar observation working independently.

Method for Mucin (After Hicks and Matthaei, 1958)

(1)   Fixation is not critical.

(2)   Bring sections to water.

(3)   Treat with 5 per cent iron alum for 10 minutes.

(4)   Wash in water.

(5)   Stain in 0.1 per cent aqueous acridine orange for 2—3 minutes.

(6)   Wash in water.

(7)   Mount in buffered glycerin (9 parts glycerin, 1 part 19/15 pG 6 phosphate buffer).

Results

Mucin fluoresces orange red, most other tissue components give dull background fluorescence.

Examine slides using BG 12 exciter filter, and OG 4 and/or OG 5 barrier filters.

Method for Fungi (Chick, 1961)

The use of iron alum as a mordant for acridine orange has been utilized in this method, which is similar to that for mucin. The fluorescent colour of the various fungi has not aided in identification; it is recommended as a simple method for the morphological identification of fungi. A variety of fungi were demonstrated by the author: *Aspergilli, Blastomyces, Coccidioides, Cryptococcus neoformans, Histoplasma capsulatum, Rhinosporidium seeberi,* in addition to actinomycotic and maduromycotic granules. A mixture of potassium hydroxide and acridine orange is recommended for direct examination of skin scrapings and hair for fungi.

Method

(1) Fixation is not critical.
(2) Stain in Weigerts iron haematoxylin for 5 minutes.
(3) Wash in tap-water for 3 minutes.
(4) Stain in 0.1 per cent acridine orange for 2 minutes.
(5) Wash in tap-water for 30 seconds.
(6) Dehydrate, clear and mount in Gurr's Xam or Fluormount.

Results

Fungi and mucin fluoresce green—red.

Examine using a BG 12 exciter filter and a yellow (OG 4) and/or orange (OG 5) barrier filter.

Fluorescent Methods for Lipids

Fluorescent methods for lipids are much more sensitive than conventional methods, and therefore require some experience in interpretation. Popper (1944) described a method using phosphine 3R which, while not as sensitive as the 3:4 benzpyrene method, has some degree of permanency. The benzpyrene method is recommended for the demonstration of the finest lipid granules, but it must be remembered that the fluorescence fades rapidly.

Phosphine 3 R Method (Popper, 1944)

(1)   Formalin fixation is preferred.
(2)   Cut frozen or cryostat sections.
(3)   Wash sections or smears in distilled water.
(4)   Stain in 0.1 per cent aqueous phosphine 3 R for 3 minutes.
(5)   Rinse quickly in water.
(6)   Mount in 90 per cent glycerin.

Results

All lipids, with the exception of fatty acids, soaps and cholesterol give a silvery-white fluorescence. See benzpyrene method for recommended filter system.

3:4 Benzpyrene Method (Berg, 1951)

Staining solution. — Prepare a saturated aqueous solution of caffeine (about 1.5 per cent) at room temperature and leave overnight. Filter, and add 0.002 g 3:4 benzpyrene to 100 ml of filtrate. Incubate at 37°C for 2 days, filter and add an equal volume of distilled water.

Method

(1)   Formalin fixation is preferred.
(2)   Cut frozen or cryostat sections.
(3)   Rinse sections or smears in distilled water.
(4)   Filter staining solution on to smears or sections and leave for 20 minutes.
(5)   Rinse in distilled water.
(6)   Mount in distilled water and examine.

Results

Lipids, even the finest granules, give a brilliant blue—white fluorescence which fades rapidly.

Examine using a UG 1 or UG 2 (BG 12 if not available) exciter filter and a colourless U.V. barrier filter.

FLUORESCENT STAINING TECHNIQUES

Fluorescent Stains for Mucin

The best fluorescent method for the demonstration of mucin is probably the fluorescent P.A.S. technique *(see below);* the acridine orange method (page 54) tends to fade on repeated examination. Vassar and Culling (1959) used 1 per cent aqueous atebrine in pH 3.95 sodium acetate–hydrochloric acid buffer for 10 minutes, followed by a brief rinse in water. This causes mucin to fluoresce a bright yellow colour, with other tissue components a pale green.

Fluorescent Feulgen Reaction (Culling and Vassar, 1961)

This technique utilizes a fluorescent Schiff reagent. It is simple in operation and, because of the intense brilliance of the fluorescence against a dark background, gives results superior to the conventional Feulgen reaction.

Control sections, treated with deoxyribonuclease (DNAse) to remove DNA, or bisulphite to block aldehyde groups, failed to stain by this technique which may be accepted as proof of specificity.

In addition to its use for the demonstration of DNA and nuclear patterns (including chromosomes), it has been utilized for the demonstration of the LE cell phenomena (Wignall, Culling and Vassar, 1962). The altered DNA of the inclusion body. which is thought to consist of DNA and histone in salt linkage, fluoresces a lighter yellow than nuclear DNA and is thus easily seen even with high dry objectives *(see Figure 5.5).* Cytomegalic inclusions may also be seen.

Fixation

Carnoy or formalin fixed paraffin sections give excellent results, as do methyl alcohol fixed smears; other fixatives may require different times of hydrolysis as for conventional Feulgen reaction.

Special Reagents Required

Fluorescent Schiff reagent

Acriflavine hydrochloride	1 g
Potassium metabisulphite	2 g
Distilled water	200 ml
N/1 hydrochloric acid	20 ml

Dissolve the acriflavine and metabisulphite in the distilled water, then add the hydrochloric acid. This should be kept overnight before use. This reagent is reasonably stable.

Method

(1) Bring sections to water.

(2) Treat sections (or smears) in preheated N/1 hydrochloric acid at 60°C for 10 minutes (depending on fixation, *see above*).

(3) Wash briefly in distilled water.

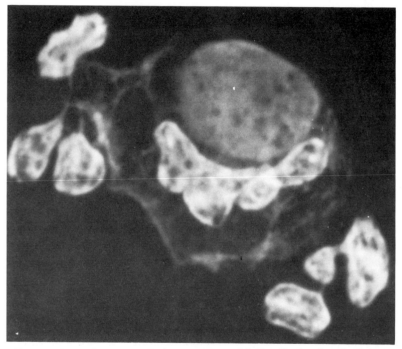

Figure 5.5 – Fluorescent Feulgen staining of L.E. cell, showing the lighter staining of the phagocytosed DNA in the cytoplasm

(4) Transfer to fluorescent Schiff reagent for 20 minutes.

(5) Wash in acid–alcohol (1 per cent HCl in 95 per cent alcohol) and leave for 5 minutes; this removes unreacted Schiff reagent and takes the place of sulphite rinses in the conventional method.

(6) Transfer to fresh acid alcohol for a further 10 minutes.

(7) Wash in absolute alcohol, a few changes to remove traces of acid.

(8) Clear in xylol and mount in H.S.R. or D.P.X.

Results

DNA . . . . Fluoresces a bright
golden yellow

Other tissue components . . . . Green

Examine using BG 12 exciter filter and yellow (OG 4) and/or orange (OG 5) barrier filters.

Fluorescent P.A.S. (Culling and Vassar, 1961)

The fluorescent P.A.S. reaction has the advantage of demonstrating minute quantities of reactive material. It demonstrates basement membranes, mucin and fungi *(Figure 5.6)* exceptionally well; it has a high degree of specificity and may be controlled in the same manner as the conventional method. Because of the degree of specificity and the brilliance of the fluorescence, some experience of the method is required in interpreting results, as compared with the conventional technique.

Fixation

As for P.A.S. technique.

Method

(1) Bring sections to water.
(2) Treat with 1 per cent aqueous periodic acid for 10 minutes. Steps 3–8 and method of examination are as for the fluorescent Feulgen technique *(see above)*.

Results

P.A.S. positive structures . . . . Fluoresce bright
golden yellow

Other tissue components . . . . Green

Demonstration of Acid-fast Bacilli in Sections and Smears

Fluorescence microscopy for the detection of acid-fast bacilli has been used widely for years. It has probably not become universally used due to the lack of, or inadequacy of, fluorescence equipment. However,

59

with the equipment now available the fluorescence method is reliable, sensitive, and permits very rapid screening of sections and smears. By using the method in duplicate with Ziehl–Neelsen technique, organisms have on several occasions been found reasonably quickly, which could only be found on repeated examinations by the conventional method.

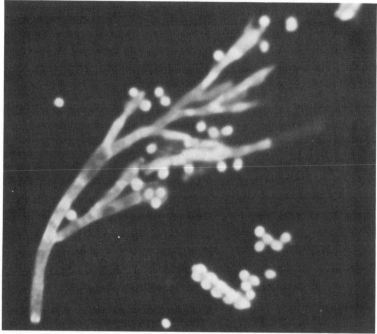

Figure 5.6 – Culture of fungi from lung, stained by the fluorescent P.A.S. technique

Wellman and Teng (1962) found that positive cases are three times as likely to be overlooked by the Z.N. method as they are by the fluorescence method.

Fixation

This does not appear to be critical. Formalin or Zenker fixatives give good results.

FLUORESCENT STAINING TECHNIQUES

Special Reagents Required

Staining solution

Auramine O	1.5 g
Rhodamine B	0.75 g
Glycerol	75 ml
Phenol cryst. (liquified at 50°C)	10 ml
Distilled water	50 ml

Method (Kuper and May, 1960)

(1) Bring sections or smears to water* (use thin, scratch-free slices).

(2) Stain with filtered auramine-rhodamine at 60°C for 10 minutes.

(3) Wash in tap-water for 2 minutes.

(4) Differentiate in 0.5 per cent aqueous HCl in 70 per cent alcohol for 2 minutes. Use 0.5 per cent aqueous HCl for *Myco. leprae.*

(5) Wash in tap-water for 2 minutes.

(6) Differentiate in 0.5 per cent potassium permanganate for 2 minutes. This step quenches background fluorescence.

(7) Wash in tap-water for 2 minutes, blot dry.

(8) Dehydrate, clear and mount in Fluormount†.

Examine using a high dry objective, with a UG 1 or 2 exciter filter, and a colourless U.V. barrier filter.

Fluorescent Technique for Alkaline Phosphatase

This method, described by Burstone (1960) as one of several new techniques for this enzyme, is one of the first applications of fluorescence to enzyme histochemistry. The enzyme releases a fluorescent naphthol compound from the substrate in a non-coupling reaction.

This approach has great significance since either the use of a substrate that gives a fluorescent reaction product, or the post-coupling of

---

*By using 30 per cent vegetable oil in xylol to remove paraffin wax from sections bacilli are more deeply stained. This step is essential when staining for *Myco. leprae.*

†This step is omitted when staining for *Myco. leprae.*

the reaction product with a fluorchrome would allow, because of the increased sensitivity of fluorescence, visualization of very low levels of activity.

Special Reagents Required

Substrate. — Approximately 5 mg 5,6,7,8,-β-tetralol carboxylic acid-β-naphthylamine phosphate and 0.5 ml *N,N*-dimethylformamide (DMF) substrate are placed in a 50 ml flask.

25 ml of distilled water is added, followed by 25 ml 'tris' buffer pH 8.7 (24.2g tris (hydroxy methyl) aminomethane, 16.5 ml N/1 HCl and distilled water to make 1 litre). Two drops of 10 per cent magnesium chloride are then added, the solution is shaken several times and then filtered. The solution should be clear or slightly opalescent.

Method

(1) Bring frozen dried, cryostat cut, or acetone-fixed paraffin embedded sections to distilled water.

(2) Incubate in substrate in a Coplin jar at 60°C for 15 minutes, then remove Coplin jar to bench for remainder of incubation at room temperature. Incubation period may vary from 1 to 3 hours.

(3) Wash slides in two changes of 50 per cent alcohol, then in running tap-water.

(4) Mount in 90 per cent glycerin.

Examine using BG 12 or UG 1 or 2 exciter filter and colourless U.V. barrier filter.

Results

Sites of enzyme activity . . . . . Brilliant bluish-white fluorescence

Fluorescent Acridine Orange Technique (Bertalanffy)

This method gives good differentiation of RNA and DNA although there is doubt as to its absolute specificity. It gives a brilliant red staining of RNA and is excellent for plasma cells and those cells actively synthesizing protein.

Solutions required

(1) M/15 Phosphate buffer pH 6.0.

(2) Acridine orange 0.01 per cent in phosphate buffer, pH 6.

(3) M/1 Calcium chloride differentiator (11.099 g in 100 ml distilled water).

Methods

Smears are fixed in ether/alcohol for at least 30 minutes. Tissue sections fixed in an alcoholic fixative (formalin fixed tissue cannot be used) are brought to water.

(1) Hydrate by passing them through 80 per cent, 70 per cent and 50 per cent alcohol for 10 seconds each, and rinse in distilled water.

(2) Treat with 1 per cent acetic acid for 6 seconds, followed by rinsing in two changes of distilled water.

(3) Stain in 0.1 per cent acridine orange for 3 minutes.

(4) Wash in M/15 phosphate buffer, pH 6.0, for 1 minute.

(5) Differentiate in M/10 calcium chloride for 30 seconds.

(6) Mount in a drop of pH 6.0 phosphate buffer, and examine under a fluorescence microscope.

Results

DNA fluoresces . . . . Green

RNA fluoresces . . . . Red

Chapter 6

Fluorescent Antibody Techniques

Creech and Jones (1940) showed that various proteins could be labelled with a fluorescent dye without material effect on their biological or immunological properties. The resultant fluorescent complex gave an intense blue fluorescence, but was difficult to distinguish from the blue autofluorescence of tissue.

Fluorochromes

In 1941, Coons, Creech and Jones described the preparation of fluorescein isocyanate (FIC) which imparted an apple-green fluorescence to the tagged antibody which could be readily distinguished from tissue autofluorescence. Although this originally had to be prepared in the laboratory, it has been available commercially since 1958, when Riggs and his colleagues incorporated a thiol group, it has been replaced by fluorescein isothiocyanate (FITC) *(Figure 6.1)* which is cheaper, more stable, and gives a more intense fluorescence.

Another fluorescent label for protein now available is 1-dimethyl-aminonaphthalene-5-sulphonic acid (DANS). This also gives an apple-green fluorescence and may be used in place of FITC. It is available commercially in a solid form*, and is cheaper than FITC. The author has had no experience with this compound, although Nairn (1962) reports that it is possible to introduce more fluorescent groups per molecule of protein with DANS (10) than with FITC (7). Most of the reported work has been done with FITC, which gives consistent results. Dyes of the rhodamine series have also been used as fluorescent antibody labels, their main area of usefulness being their orange-red colour

*Fluka, Buchs, Switzerland.

FLUORESCENT ANTIBODY TECHNIQUES

which contrasts well with FITC. Rhodamine B isothiocyanate (Riggs and colleagues, 1958) and tetramethylrhodamine isothiocyanate (Hiramoto and colleagues, 1958), the most commonly employed for this purpose, may be used in double tracing experiments or, after conjugating with albumen, as a non-specific counterstain.

It is thought that these fluorochromes attach to proteins by the ε-amino groups of lysine (Nairn, 1964) by a thiocarbamide bond (Kawamura, 1969). Since it is known that a molecule of immunoglobulin G (IgG) contains 86 lysine residues, and that the net negative charge of a protein is increased for each molecule of dye attached, labelling of a large number of lysine residues would result in non-specific staining, Schiller and colleagues (1953) and Curtain (1958) found that the change in net charge was consistent with an average of 1–2 molecules of fluorochrome per molecule of protein.

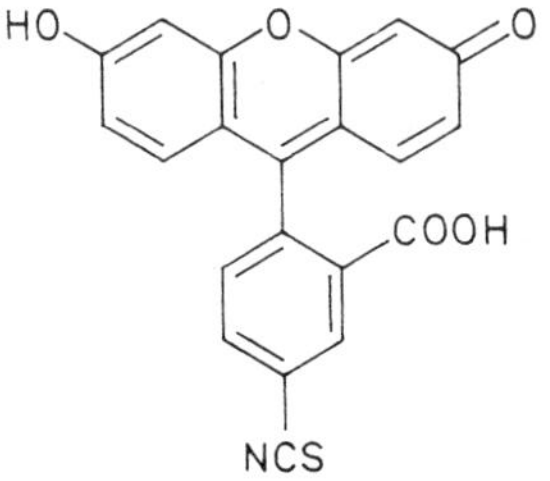

Figure 6.1 – Fluorescein isothiocynate

Inactivity of Antibody

The estimates of the reduction of antibody activity following conjugation with fluorochromes range from 50 per cent (Coons and Kaplan, 1950; Chadwick and colleagues, 1958) to nil (Mayersback, 1958; Redetzki, 1958), from the evidence available it would seem to be of the order of 25–30 per cent.

Methods of Use

Antibodies are produced by a series of injections of antigen (with or without an adjuvant) into an appropriate animal.

Freund's adjuvant was, until recently, widely used as an adjuvant to stimulate the immune system, but it has the disadvantage of causing excessive production of non-specific antibodies. More recently, aluminium–potassium sulphate has been used as an adjuvant, which gives a good antibody titre and fewer non-specific antibodies. Kawamura (1969) recommends a ration of protein to alum of 1:256 using 1 vol. of

66

1 per cent antigen protein to 25.6 vols. of 10 per cent alum buffered to pH 5.5 with phosphate buffer. With the addition of the alum to the protein a precipitate is formed, this is centrifuged and the precipitate resuspended in saline solution at pH 5.5. Alternatively serum from a patient or animal with naturally occurring antibodies may be used; the gamma globulin fraction is usually then precipitated, to which the fluorescent dye is coupled (conjugated). This conjugate may then be used to detect and locate antigen (or antibody) by one of the following methods.

Direct Staining

Here the fluorescent antibody is used to directly locate an antigen. The section or smear is flooded with the conjugate, left to react, then washed with buffered saline to remove unreacted antibody. The presence of apple-green fluorescence will indicate the location of antigen. *Figure 6.2* illustrates this method.

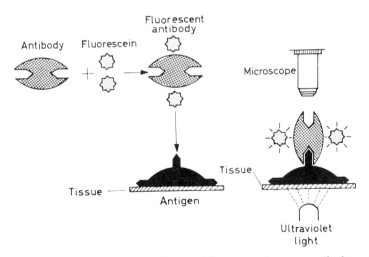

Figure 6.2 – Direct staining. Attachment of fluorescent dye to an antibody, and its subsequent attachment to an antigen in the specimen

Sandwich Technique (Weller and Coons, 1954)

By this technique, antibody sites may be visualized. The smear or section is first treated with unlabelled antigen which will attach itself to antibody. After an appropriate time this is washed off and replaced by a fluorescent tagged antibody. The tagged antibody will react (or bind) with the antigen which has attached itself in the first stage to the antibody in the original specimen, thus making a sandwich. The presence and location of positive fluorescence now indicates antibody sites in the original specimen *(Figure 6.3)*. Sandwich staining is more sensitive than simple staining. Coons (1956) has assessed the increase in sensitivity as tenfold.

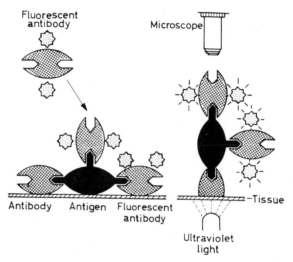

Figure 6.3 – Sandwich technique. Following attachment of an unlabelled antigen to an antibody in the specimen, a fluorescent labelled antibody is attached to the antigen

Multiple Layer Technique

This is an extension of the sandwich technique. This technique for antigen (or antibody) can be made more sensitive by successive layering of antibodies; for example, if mouse antibody (globulin) is to be detected, the specimen is as follows.

(1)   Treated with unlabelled rabbit anti-mouse gamma globulin. This will combine with any mouse globulin that may be present. Uncombined antisera is washed off.

(2)   Treated with unlabelled goat anti-rabbit sera which will combine with the rabbit globulin used in Step 1.

(3)   Theoretically one can now use alternately rabbit anti-goat,

and goat anti-rabbit and build a bigger and bigger aggregate around each molecule of the original mouse antibody. It is generally assumed that there are several reactive sites on each molecule of antibody. Therefore one molecule of antibody (x) will bind x number of rabbit anti-mouse molecules (stage 1) and there will be x^2 after Stage 2, and x^3 after Stage 3, and so on. The last antibody used will, of course, be labelled.

It will be found that in practice there are a great number of technical problems involved in such a procedure, not the least of which is removing unreacted antibody.

In Vivo Tracing

Pressman, Yagi and Hiramoto (1958) studied the *in vivo* localization of anti-tissue antibodies by this fluorescent antibody technique. He concluded that the *in vivo* technique was at least 4–12 times less sensitive. *Figure 6.4* shows a section of kidney from a rat injected with rabbit (anti-rat glomerular basement membrane) sera which has been stained with labelled goat anti-rabbit gamma globulin.

Application

The fluorescent antibody technique has been used for the detection and localization of bacterial, viral, protozoal, fungal, helminthic, and animal and human tissue antigens (Nairn, 1964; Kawamura, 1969). It has also been used, as described above, for the detection and localization of antibody (Coons, Leduc and Connolly, 1955). By absorption and cross absorption techniques, it has the specificity of immunological techniques and has a sensitivity and precision that enables identification and localization at a microscopical level. In the diagnostic laboratory it is becoming increasingly widely used in the bacteriological field. The field of auto-immunity, in particular, has received great stimulus by this technique since fluorescent labelled patient's serum can be seen to react with various components of the patient's own tissue. The demonstration of anti-nuclear antibodies in the sera of patients with various collagen diseases (Fennell, Rodnam and Vazquez, 1962); the presence of gamma globulin concentrations in the lesions of various diseases (Vazquez and Dixon, 1957); the demonstration of rheumatoid factor by Mellors and colleagues (1961), and Taylor and Shepherd (1960); and the formation of antibodies to thyroglobulin in immune thyroiditis by White (1957) are but a few examples of its use.

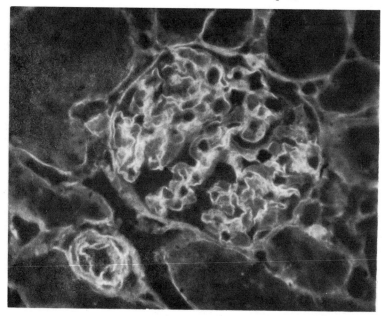

Figure 6.4 – Direct staining. Rat kidney stained with anti-basement membrane conjugate. The basement membranes fluoresce bright apple-green

Non-specific Staining

Staining of a specimen which is not due to a specific reaction between a particular antigen and the labelled specific antibody is referred to as non-specific staining. In spite of many improvements in technique this still presents a serious problem, particularly when dealing with tissue antigens and antibodies.

Unreacted Fluorescent Material (UFM)

The original method for the removal of UFM (unattached FITC) from conjugated sera, was by simple dialysis; this was never completely satisfactory. This problem has now been overcome by the use of gel filtration with Sephadex, or by extraction with activated charcoal.

Conjugated Normal Serum Proteins

These are proteins, other than antibody proteins, which have been labelled during the conjugation process. Because of their non-specificity

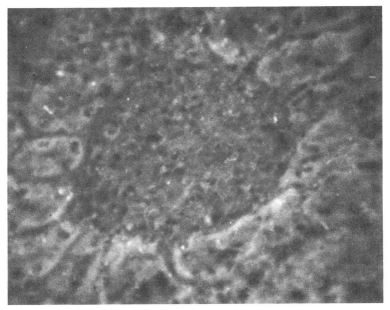

Figure 6.5 – Specific absorption. Rat kidney stained with anti-basement membrane conjugate that has been absorbed twice with antigen (basement membrane suspension, see Figure 6.4)

they attach themselves arbitrarily to protein in the specimen, giving the appearance of a specific reaction. Coons and Kaplan (1950) used dried tissue powders to selectively remove them. The powder is preferably prepared from the same type as the tissue to be stained. The absorbing powder should, of course, not contain the specific antigen to be detected; if the antigen is present in the powder to any significant degree it will remove the conjugated antibody. Fractionation of the serum and conjugation of only the gamma globulin is a valuable practical method of reducing this type of non-specific staining; however, this conjugate should still be absorbed with tissue powder.

Conjugated Unwanted Antibodies

These are antibodies, present in the serum to be conjugated, which have been produced either as a result of natural infection, by impurities in the immunizing antigen, or are against organisms or tissue components which share a common antigen with the material to be investigated (Coons, 1951). Such antibodies must be removed. The sera are absorbed with appropriate antigens, for example, similar tissue or strain of organism which does not contain the specific antigen being traced.

Conjugation

This is the process of coupling the fluorochrome to protein. The antisera (or protein antigen) may be as follows.

(1) Conjugated as whole sera.
(2) Fractionated after conjugation (as 1) and only the gamma globulin fraction used.
(3) Fractionated to obtain gamma globulin fraction, and only the globulin fraction conjugated.

Method (1) is not recommended for antisera since it will contain a large amount of conjugated serum proteins which will increase the amount of non-specific staining. Methods (2) and (3) are those in general use; the author uses the latter since this economizes in the amount of FITC used.

Fractionation of Sera

Antibody present in sera is concentrated by precipitating the gamma globulin present. To each ml of sera is added 1 ml of saline and 1.33 ml of cold saturated ammonium sulphate at 4–6°C. After centrifugation, the precipitate is washed once with 50 per cent saturated ammonium sulphate and dissolved in buffered saline. The globulin suspension is then dialysed overnight against buffered saline. The protein concentration is determined by the biuret reaction, and the concentration adjusted to 50–60 mg/ml. It should be not less than 25 mg/ml (Goldstein and colleagues, 1961). This solution can then be stored in a deep freeze.

Conjugation

To one volume of globulin add two volumes of cold 0.5 M carbonate–bicarbonate buffer pH 9.0*. This solution is stirred thoroughly at a temperature of 0–2°C (in a cold room or ice bath) while 1 mg of FITC dissolved in the minimum amount of acetone is added for each 18–20 mg of protein. The acetone solution is slowly

*pH 9.0 buffer. 3.7 g $NaHCO_3$ and 0.6 g Na_2CO_3 (anhyd.) are dissolved in distilled water, and made up to 100 ml.

added over a period of 15 minutes. The stirring is continued overnight. Care should be taken to ensure that the stirring, while thorough, does not cause frothing.

Purification of Conjugates

Unreacted fluorescent material (UFM) was originally removed by dialysis against saline over a period of several days, but this method is not as effective as the charcoal or Sephadex methods.

Extraction with powdered activated charcoal, which has been washed well with saline and dried at $100°C$, is much more effective than using untreated charcoal. The charcoal, moistened with saline to avoid undue loss of protein, is added in the proportion of 2.5 mg/mg of protein in the sera. The mixture is shaken for one hour and the charcoal is then removed by centrifugation. The only disadvantage of this technique is the loss of protein (20–30 per cent) which may be important when conjugating small amounts of a weak antisera.

Gel filtration with a cross-linked dextran, Sephadex* is the better and more popular method, it depends on the diffusion of small molecules into the pores of the gel, the larger molecules being excluded because of their size. Separation takes place in a column, with the large protein molecules travelling more rapidly than the small molecules, which diffuse into the gel. Pore size G 25 or G 50 are the commonly used grades for this purpose. The manufacturers describe the method of preparation and use in detail. For volumes of conjugate up to 20 ml a column 20 cm in length and 3 cm in diameter, which has been washed with buffered saline, is adequate. The conjugate having been centrifuged, is allowed to soak into the column, and a suitable head of buffered saline applied. As the solution passes down the column, two bands separate, the faster one being the conjugate. The loss of protein by this method is very small. Dilution of the conjugate may be overcome by reprecipitation of the globulins.

Absorption by Tissue Powders

The removal of conjugated non-specific serum proteins *(see above)* is usually carried out by absorption with tissue powders, if possible with the same type as the tissue to be examined, provided that it does not contain any appreciable amount of the specific antigen. The precipitation of the globulin fraction will remove these to a great degree, but

*Pharmacia Ltd., Uppsala, Sweden.

more elaborate fractionation of sera by chromatography on modified cellulose (Goldstein and colleagues, 1961) reduces them to almost undetectable levels.

The tissue powders, usually liver or bone marrow (to inhibit non-specific staining of granulocytes) are prepared as follows.

Wash the organ free of blood with physiological saline, chop into small pieces with scissors or scalpels, and rewash with saline. Grind up the material in a low-speed homogenizer (or pestle and mortar) with acetone, and filter through coarse filter paper. Wash several times with acetone until completely dehydrated, and dry at 37°C. Grind to powder in a mortar, sieve through wire mesh to remove coarse material and store at room temperature. For absorption purposes approximately 100 mg of tissue powder is used for each ml of original serum. The mixture is shaken at room temperature for 1 hour, and centrifuged at about 10,000 g (preferably in a refrigerated centrifuge) for about 15 minutes. The high speed is essential to give maximum return of conjugate. The supernate is now ready for use. The conjugate should be divided in small aliquots and stored in a deep freeze, otherwise there will be a protein–dye breakdown which will necessitate re-absorption before use.

Prepared Conjugates

Commercially prepared conjugates are available which are ready for immediate use. They are commonly used for the demonstration of specific 'antigens' such as the immunoglobulins IgG, IgM, IgA, and so on, and the heavy or light chains, of which the immunoglobulins are composed and in addition for complement (β1c), bacteria, and so on. Conjugated antisera to mouse, guinea-pig, rabbit, sheep sera (or IgG) are available for use in sandwich techniques which makes conjugation of specially prepared antisera optional, for example, antisera prepared in a rabbit can be demonstrated in a treated preparation by the use of conjugated sheep anti-rabbit sera.

Preparation of Material to be Stained

Cryostat Sections

These are by far the most commonly employed histological preparations. Unfixed tissue is quick frozen or quenched and sections cut in the cryostat. They may be used fresh (after air-drying) or fixed in cold 95 per cent alcohol either before or after air drying. Tissues which tend to detach during staining may need to be air-dried before fixing. The type of fixative used will depend on the antigen/antibody involved in the reaction, but cold 95 per cent alcohol has been employed successfully in a number of investigations.

Freeze-dried Sections

These have been used with success (Rey, 1960). Tissues after drying are embedded in polyester wax.

Paraffin Sections

Methods have been described using freeze-substitution techniques but they are still at an experimental stage.

Conventional paraffin sections have been used on a few occasions but are generally unsuitable.

Smears

Tissue and bacterial smears, touch preparations and tissue culture monolayers on coverslips have all been used with great success for fluorescent antibody staining techniques.

Smear preparations of tissue are best made by brushing the cut surface of the tissue with a camel hair brush which has been dipped into 7.5 per cent PVP (polyvinyl pyrollidone). Several strokes are then made on to a clean slide. One advantage of this technique is that the experimental and control smears can easily be made on the same slide.

Staining Technique

The method of mounting the stained slides recommended is that described by Culling (1967), using New Unimount. Alternate slides may, sometimes, with advantage be examined in buffered glycerin.

Reagents Required

Phosphate buffered saline (pH 7.1)

Sodium chloride . . . . 8.5 g
Disodium hydrogen phosphate (anhyd.)
  (Na_2HPO_4) . . . . 1.07 g
Sodium dihydrogen phosphate
  ($NaH_2PO_4 . 2H_2O$) . . . . 0.39 g
Distilled water . . . . to 1 litre

FLUORESCENT ANTIBODY TECHNIQUES

Buffered glycerin mountant

Glycerin . . . . 9 ml
Phosphate buffered saline . . . . 1 ml

(1) Sections or smears may be rinsed with buffered saline. This facilitates spreading of the conjugate.

(2) Slides or coverslips are placed in a moist chamber, for example, Petri dish, with moist filter paper in the bottom. The preparation is covered with a drop (or two) of the conjugate which is applied with a platinum loop or Pasteur pipette. The chamber is kept at room temperature. The reaction time may vary (with the strength of antisera or type of antigen) from 10 minutes to 2 hours; from 15 to 30 minutes is usually adequate.

(3) The preparations are rinsed in several changes (not less than three) of buffered saline over a period of 10–15 minutes. This step is critical and should be carried out for a longer, rather than a shorter time. Dr. Irene Batty of the Wellcome Research Laboratories uses a magnetic stirrer to agitate the buffer during washing, her stained preparations were quite the best I have seen and we now use and recommend this method. The slides to be washed are suspended in an open-type slide holder in a beaker, over a magnetic stirrer; three changes are usually adequate.

(4) Excess buffered saline is wiped off and the specimens mounted in buffered glycerin. After examination they may be rinsed in distilled water and stored dry in the dark.

or

(4) Rinse briefly in distilled water and blot dry.

(5) Place in xylol until section is clear; it may be necessary occasionally to blot again before complete clearing is achieved.

(6) Mount in New Unimount*.

Results

Antigen–antibody reaction sites give an apple-green fluorescence (with FITC or DANS) or orange fluorescence with rhodamine B.

Background fluorescence will be blue (unless counterstained with rhodamine B) except for autofluorescent sites (*see* page 52). An unstained slide should be used to check autofluorescence.

*Obtainable from Sherwood Medical Industries, Inc. 1831 Olive St., St. Louis, Missouri, U.S.A. 63103.

Examine using a Schott BG 12 (alone or preferably with a Schott UG 1) exciter filter and one of the following barrier filters; Euphos, yellow green GG 9, Zeiss 44 or their equivalent.

Tests of Specificity

Tests of specificity are modelled on those used in established immunological techniques. Those most commonly employed are as follows.

Blocking

Staining should be inhibited by pre-treatment of the specimen with unconjugated antisera (at room temperature for 30 minutes) before staining. The unconjugated sera should bind all the reactive sites on the antigen, and block a reaction with the conjugated antibody. This is referred to as a *blocking test*. It will sometimes be found that only a reduction in intensity of staining can be obtained by this test.

The blocking test should also be performed with a non-specific (control) serum to prove that the blocking is due to the presence of the specific antibody.

Absorption

Staining should be inhibited if the conjugate has been previously absorbed (usually twice) with the specific antigen (*specific absorption*), the conjugate being centrifuged after absorption to remove reacted material (if possible).

The staining should not be inhibited by absorption with a different antigen *(non-specific absorption)* using the same technique.

Absorption of the antisera with an antigen is simple when the antigen is pure, but it must be remembered that when an antigen is impure there is no guarantee that positive staining of the antibody may not be due to a non-specific antigen (impurity) blocking a non-specific antibody.

Control Conjugate

Occasionally it may be necessary to prove that staining is due to an induced antibody, and not a naturally occurring one. This is tested for by conjugation of a normal control serum from the same

FLUORESCENT ANTIBODY TECHNIQUES

Treatment (as detailed above)	Result if sera is specific
(1) Conjugated antisera alone	Staining (+ + +)
(2) Unconjugated antisera, followed by conjugated antisera. (Blocking test)	No staining (−)
(3) Unconjugated control sera, followed by conjugated antisera. (Control blocking test)	Staining (+ + +)
(4) No treatment. Slide mounted unstained to control autofluorescence	No staining (−)

species of animal. There should be no staining of the antigen with this sera.

Routine Test and Control Technique

As a routine we set up four slides of each specimen to be examined. The treatment and expected results are shown above.

Chapter 7

The Polarizing
Microscope

Theoretical Aspects

Although it is beyond the scope of this book to enter deeply into the
theory of polarized light, the basic elements of the subject will be
explained.

Light is assumed to be due to a wave motion, to the upward and
downward vibration of ether particles. These do not move along in the
direction of the light ray (*Figure 7.1,* A to B), but vibrate at right angles
to it (*Figure 7.1,* C to E), and when the light ray ceases they return to
their original position (*Figure 7.1,* D). Ether is supposedly an homogeneous medium and there is no reason therefore to believe that these
particles will vibrate in any one direction more than another. To
explain polarized light it is necessary to suppose that light normally
vibrates in all planes; that is, in *Figure 7.2,* from C to E, F to G, H to I,
and J to K. It is difficult to imagine a particle oscillating in all planes at
one time, but it is possible to imagine that it moves – at right angles to
the direction of the light ray – in all planes in such rapid succession so
as to act as if it were moving in all these planes at one time.

While this theory is not strictly accurate, it is sufficiently correct to
explain the behaviour of polarized light.

If a dot drawn on a sheet of white card is viewed through a block of
glass laid on top, only one dot will be visible from above *(Figure 7.3a).*
If the block of glass is replaced by a polished block of crystal, such as
Iceland spar, two dots will be visible. Such a crystal is described as
being bi-refringent or anisotropic; it has split each light ray from the
dot into two rays which emerge from the crystal at different points
(Figure 7.3b).

This splitting of light rays by certain crystals is due to their uneven optical density. It is known that light rays are retarded when travelling through an optically dense medium such as glass, but since the molecules in glass are evenly spaced in all directions only a simple retardation or slowing takes place. The molecular structure of a crystal differs from glass in that although its molecules are regularly spaced they are closer

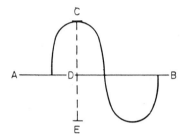

Figure 7.1 – Sine curve representing a wave of light
AB = direction of travel
CDE = direction of vibration

together in one direction than in another; they are therefore unevenly dense. There are many types of crystalline structure, but all have the common property of being more dense in one direction than in another.

Figure 7.2 – Theoretical representation of the vibration of light in all planes

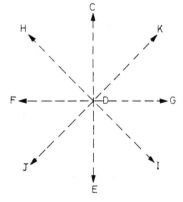

Figure 7.4 represents a downward view of a series of posts through which wind is blowing from directions W1 and W2. It will follow that from whatever angle the wind is blowing it can only leave in the direction of A or B. The intensity of the wind emerging at points A or B will depend on the angle at which it enters. If it enters from W1 then almost all the wind will emerge in direction B; if the wind enters from W2, an almost equal amount will emerge in each direction. If we now substitute the words 'crystalline structure' for 'posts driven into the ground',

and 'light rays' for 'wind', an understanding may be gained of what happens when a light ray passes through a crystal: the ether particles are vibrating in all directions at right angles to the line of propagation when it enters, but two rays emerge, and each of these causes ether particles to vibrate in one plane only (A to C, or B to D in *Figure 7.4,* these two planes always being at right angles to each other).

Light when entering a dense medium is retarded in speed. Further, being an unevenly dense medium, the crystal will retard the two rays to a differing degree, and since refraction is partly dependent on density, the two rays will be refracted or bent to differing degrees. This is known as double refraction or birefringence and explains the phenomena described above *(Figure 7.3)*. A ray of light entering such a crystal will be converted into two rays *(Figure 7.3c,* B and C) which will emerge at different points, and the emergent light rays will be polarized; that is, all the vibrations in one ray (B) will be in one single direction (B_1-B_2); in the other ray (C) in another single direction (C_1-C_2), and these directions will be at right angles to each other.

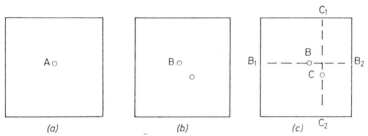

Figure 7.3 — Optical effect of (a) block of glass and (b and c) birefringent material when laid on a single dot

The Nicol Prism

Just over 100 years ago, Nicol devised a prism from which light rays, having passed through, would emerge vibrating in a single plane, that is, as polarized light. The single direction in which the light is vibrating when it emerges is known as the 'optical path' of the prism.

The prism is composed of a crystal of Iceland spar, cut to the shape shown in *Figure 7.5,* slit in half and the halves cemented together with Canada balsam along the line CB–CB. On entering the prism, a light ray (A) is divided into two rays (B and C) which are refracted differently, ray C being refracted to one side. Owing the the difference in the refractive index between Canada balsam and the calcite spar crystal and the cement, ray C on meeting the surface CB–CB at a greater angle than ray B, is totally reflected out of the prism. Ray B passes through the prism and emerges vibrating in the direction of the optical path of the prism only and is polarized light.

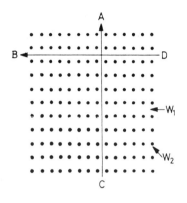

Figure 7.4 – Diagram illustrating the theoretical effect of birefringent material in splitting a beam of light (or wind) into two beams (CA, BD)

It will follow that if another Nicol prism is placed above the first one *(Figure 7.6)*, the polarized light ray B will pass through the upper prism if their optical paths are aligned (B_1), but if the upper prism is rotated

Figure 7.5 – The Nicol prism

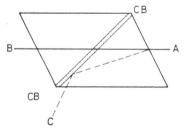

through 90 degrees so that the optical paths of the prisms are crossed, then ray B will be totally reflected out of the upper prism (B_2). Such prisms are said to be crossed and it will be seen that no light will normally emerge from crossed Nicol prisms. If the upper prism is slowly rotated it will be seen that the amount of light passing through will vary with their relative positions. At a rotation of 45 degrees, from alignment of the prisms, approximately half the light will pass through the prism, and so on. In practice, it will be found that with an intense light source some light will pass even through crossed Nicol prisms, but with light of moderate intensity the field will appear black.

Polaroid Discs

In 1935 'polaroids' — glass or celluloid covered discs with the ability to polarize light — were first made available for use in place of Nicol prisms. They act as a single crystal of herapathite which is not only birefringent, but has the ability to absorb the ordinary ray (which would be refracted out of a Nicol prism (*Figure 7.5,* C)), only the extraordinary ray (*Figure 7.5,* B) being transmitted.

Polaroids are made by suspending ultra-microscopic crystals of herapathite in nitrocellulose. All the crystals in the suspension are orientated so that their optical paths are aligned. This suspension when mounted between two glass plates or celluloid sheets acts as a single crystal.

One glass plate is made to fit into the substage filter carrier, and the other has a metal mount to hold it in place on top of the eyepiece. The celluloid sheet may be cut with scissors and used in a similar manner. For all practical purposes they may be used as Nicol prisms.

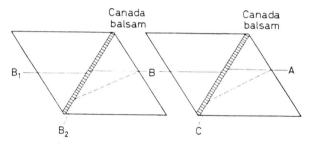

Figure 7.6 — The effects of two Nicol prisms, with optical paths aligned (ABB₁) and optical paths crossed (ABB₂)

The Use of the Polarizing Microscope

It has been shown that certain crystals have the power to convert a single ray into two rays of light, which are vibrating in a single plane at right angles to each other *(Figure 7.3)*, and also have the power of quenching or absorbing one set of these rays.

If such a crystal is placed on the stage of a microscope having a Nicol prism (or polaroid) in the substage, the effect will be as shown in *Figure 7.7*. The light rays in the field will be vibrating in the optical path of the Nicol prism (AB), except those that pass through the crystal, which will be vibrating from C to D or E to F. If the set of rays CD were absorbed (as described above for herapathite), the crystal would have the effect of changing the plane of vibration from AB to EF. Therefore, a Nicol prism placed above the object, having its optical path from A to B, will absorb most of the rays which have passed the crystal, and it will appear dark on a light background. If the upper Nicol prism has its

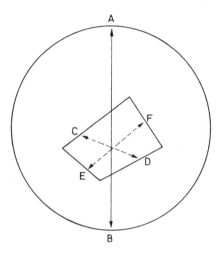

Figure 7.7 – Showing birefringent effect of a crystal

optical path in the direction EF, then the rays AB will be absorbed and the crystal will appear light on a dark background. This latter method is the one normally used in histological laboratories, since the majority of the birefringent material to be examined consists of small particles or crystals which are easier to see as light objects on a dark background. It will be seen that such crystals have the apparent power of changing the plane of vibration (from AB to EF) and for this reason they have occasionally been referred to as 'optically active'.

The direction in which the plane of vibration is changed, and the degree to which it is turned, may be used to assist in the identification of a crystal. For this reason, petrological microscopes have the polarizing prism (in the substage) in a graduated circular mount which may be rotated through 360 degrees. In addition, the stage is usually graduated and may be rotated in the same manner, so that the object may also be orientated.

Crystals which change the plane of vibration clockwise are called 'dextra-rotary', and anti-clockwise 'laevo-rotary'.

In the histological laboratory, one is usually only concerned with simple birefringence. The microscope is set up, as described on page 24, polarized discs normally being used. When the material being investigated is in the field, the upper polaroid (analyser) is rotated. The material will appear light on a dark background if it is birefringent or anisotropic. Talc crystals, hair, myelin, silica and collagen fibres are among the many birefringent substances found in histological sections.

Types of Birefringence

Certain crystals or tissue structures show more than one index of refraction for a given wavelength of light and are therefore said to be doubly refracting or birefringent. As was shown above, this is due to some sort of asymetrical and orientated spatial arrangement of particles. These particles carry resonating charges capable of interacting with the oscillations of light waves. Birefringent material may show one or more than one type of such arrangement; the more common types may be characterized as follows.

Intrinsic or crystalline birefringence. – This refers to a type of anisotropy due to an asymetrical alignment of chemical bonds, ions or molecules. Many crystals display this type of birefringence, it is also common in biological objects such as collagen and muscle fibres, and chromosomes.

Intrinsic birefringence in a specimen is independent of the refractive index of the immersion medium which is probably due to the fact that the orientated elements are of close structure between which the medium does not penetrate.

Form birefringence. – This is found in mixed bodies, wherein asymetrical particles of one refractive index are dispersed in a specially oriented manner in a medium having a different refractive index. At least one dimension of the particles must be small in relation to the wavelength of light employed. These dispersed particles may be filaments, sheets, and so on, and they may be dispersed in a liquid, gas or solid; they can give rise to birefringence even if separately either or both are isotropic. Tests for form birefringence depend upon causing media of varying refractive index to penetrate between the particles when, at the appropriate R.I., form birefringence will disappear. (Examining objects mounted in a variety of mountants with differing R.I., for example, water, glyccrol, IISR, and so on.)

Strain birefringence. – When a dielectric substance is subjected to mechanical stress, the bonds within the substance can be distorted and give rise to a pattern which will result in birefringence. This is most simply demonstrated by twisting clear plastic (Perspex) between crossed polaroids when a birefringent spectrum of colour is produced. Similarly, glass or elastic tissue fibres under stress show birefringence.

Positive and negative birefringence. – A certain amount of confusion has árisen due to the use of *the term 'negative birefringence' to indicate a type of birefringence and not its absence.* An object that appears bright on a dark field, when viewed between crossed Nicol prisms (or polaroids), is said to be birefringent *(see above).* This does not allow the determination of direction of the fast and slow axes of the doubly refracting material; having two different R.I.s one light ray will be retarded (slow) in relation to some distinguishing dimension of the object.

In a collagen fibre the slow axis of transmission is parallel to the long axis of the fibre, the fibre is thus said to show positive birefringence with respect to its long axis; conversely, a chromosome shows negative birefringence with respect to its long axis.

To determine the sign of birefringence in a fibre or crystal (for example, to differentiate between the urate crystals of gout from calcium pyrophosphate dihydrate (CPPD) one needs, in addition to a pair of Nicol prisms (or polaroids), a first order red quarter-wave plate. Focus the object to be examined between crossed polaroids (or Nicol prisms) and insert the quarter-wave plate between the object and the analyser (upper prism) and rotate the plate until its slow axis of transmission (usually indicated by a stamped line or arrow) is parallel with the fibre or long axis of the crystals (E–F in *Figure 7.7*). The field will now be uniformly red (from its interference effect) and if the fibre or crystal appears blue then the slow axes of the plate and object are parallel and this is positive birefringence, if the object appears yellow it indicates negative birefringence (the slow axis of the plate is parallel with the fast axis of the object). In *Figure 7.7* if the slow axis of the quarter-wave plate is parallel with E–F and the crystal appears blue on a red background this is positive birefringence, if it appears yellow on a red background then it displays negative birefringence (with respect to its long axis). Dichroism is also detected by use of a polarizing microscope.

Summary

An over-simplification of the theory of polarization, but one which illustrates its normal use in histology, is as follows *(Figure 7.8)*. The Nicol prisms (or polaroids) act as grilles, and allow only light

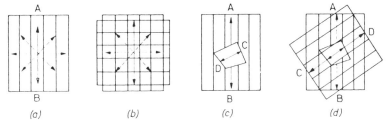

Figure 7.8 – Diagrammatic summary of the theory of polarization

vibrating in a single plane *(Figure 7.8a,* AB*)* to pass through; by inserting a further prism above the first one (*Figure 7.8b*) all light is stopped. If the light from such a prism passes through certain crystals its direction is rotated from AB to CD *(Figure 7.8c)* and a further Nicol prism placed above the crystal in the direction CD *(Figure 7.8d)* will stop all the light from the lower prism except that which has passed through the crystal and is vibrating in the optical path of the upper prism, so the crystal will be light on a black background.

Chapter 8

The Phase–Contrast Microscope

The introduction of phase-contrast microscopy is probably the greatest single advance in biological laboratories in this century. For the first time, living organisms and cells may be examined in detail without previous treatment, and the image produced a true one.

All living cells and organisms, although appearing almost homogeneous when examined unstained by ordinary microscopic methods, are composed of minute structures having slight differences in refractive index. The phase-contrast microscope, by converting these slight differences in refractive index into changes of amplitude (or brightness), produces an image that may be accurately focused and one that is easily seen or photographed.

Professor Zernicke, who was awarded the Nobel prize for his work on phase contrast, first applied his original work on telescopes to the biological microscope in 1935, but it was not until 1945 that a commercial model was available in Great Britain, although Burch and Stocks described a method for the conversion of an ordinary microscope in 1942.

These microscopes are now being widely used, and there seems little doubt that they will in the future replace what we now know as the routine microscope, since they may be used either for phase-contrast microscopy or for routine purposes.

Principles of Phase-contrast Microscopy

Without going too deeply into the theory of phase contrast, the first part of this chapter is intended to explain the broad principles underlying its use. To understand these principles it is necessary to recall some of the properties of light rays.

Light, arising from a point source, may be represented by straight lines or, since it is propagated in waves, by *sine* curves. These curves are a useful method of representation since they can be made to show not only amplitude and wavelength but the retardation of one ray in relation to another. *Figure 8.1* shows the method by which amplitude and

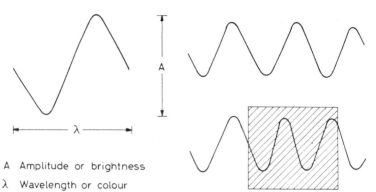

A Amplitude or brightness

λ Wavelength or colour

Figure 8.1 — Sine curve representing a ray of light   *Figure 8.2 — Retardation of lower light ray by a block of glass*

wavelength may be represented using the *sine* curve. Retardation of one light ray in relation to another is shown in *Figure 8.2* where the lower ray, having passed through a block of glass, is retarded by half a wavelength. It will be appreciated that the eye is sensitive to changes in amplitude (or brightness) and to changes in wavelength (which are changes in colour) but not to changes in phase, where one wave is retarded in relation to another.

One further property of light that must be considered is that of interference *(see Figure 8.3)*. If two rays of light strike a screen at the same point, as in *Figure 8.3a*, the resultant light on the screen will be the sum of the amplitude of the two rays (*b* and *c*) as shown in the *sine* curve (*d*). If one of the two rays (arising from the same point source) had passed through a block of glass, of an exact thickness to retard that ray by half a wavelength, instead of getting an increase in light there would be no light. This is because coherent light rays have the property in interfering with each other. This interference may (as shown in *Figure 8.3e*) result in complete extinction of the light or in a reduction of it depending on the relative amplitudes of the light rays. If, for instance, in *Figure 8.3* the direct ray in (*e*) has the amplitude of (*d*) and the retarded ray the amplitude of (*b*) instead of extinction of the light, there would have been light of the amplitude of (*c*). For practical purposes it may be said that when coherent light rays interfere the amplitude of the resultant ray can be obtained by subtracting the amplitude of one from the other. Fractional phase differences (for example, ¾ or ¼ of a wavelength) between rays will result in partial interference and in this way an image of an unstained object may be built up.

By almost closing the iris diaphragm of the substage condenser of the normal microscope an image of this small aperture is formed at the back focal plane of the objective *(Figure 8.4b)*. This image, which can be seen by removing the eyepiece, may be focused with an auxiliary eyepiece. If a phase grating is placed in the object plane (on the stage), instead of only one image being seen at the back focal plane of the objective three images are visible, the original bright one in the centre, and a further two, one at each side, which are less bright *(Figure 8.4a)*.

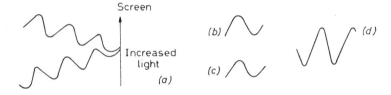

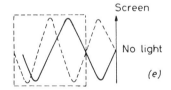

Figure 8.3 – Resultant amplitudes of light rays striking the same point on a screen, in phase (a); out of phase (e)

A phase grating consists of alternate strips of material of differing refractive index and the additional images have been formed by the light diffracted by the object. These images, known as diffraction images, may, by this method, be differentiated from the direct image. One other effect that has taken place, which is not apparent but can be proved, is that the diffraction images are out of phase with the direct rays by ¼ of a wavelength (¼λ), that is, they have been retarded by a ¼λ in relation to the direct rays *(Figure 8.4c)*.

If an annulus is placed on the substage condenser an image of that annulus will be formed in the back focal plane of the objective, and an object possessing slight non-homogeneities (such as unstained living cells) placed in the object plane will produce a halo of light both inside and outside the annular image. This halo is composed of light rays which have been diffracted by the object and are ¼λ out of phase with the direct light rays.

It will follow that if the diffracted rays of light could be retarded a further ¼λ, then the phase difference between the direct and diffracted rays would be ½λ and interference (as in *Figure 8.3e*) would take place

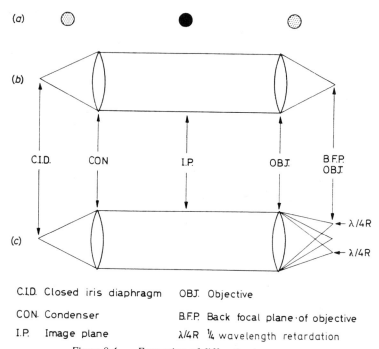

C.I.D. Closed iris diaphragm   OBJ. Objective

CON. Condenser   B.F.P. Back focal plane of objective

I.P. Image plane   λ/4R ¼ wavelength retardation

Figure 8.4 – Formation of diffraction images

in the final image plane, building up a picture, in light and shade, of an unstained specimen. Zernicke devised the 'Z' plate, now known as the phase plate, which, placed at the back focal plane of the objective, brought this about. The phase plate consists of an optically plane glass disc out of which is cut a channel to coincide with the image of the light annulus. The depth of the channel must be the exact depth to retard the diffracted rays, which travel through the full thickness of the plate, by a ¼λ in relation to the direct rays which travel through the channel *(Figure 8.5)*.

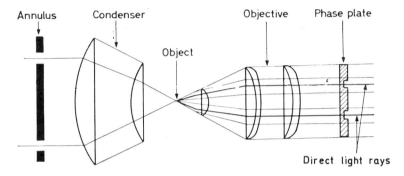

Figure 8.5 – Passage of light rays through the optical components of the phase-contrast microscope

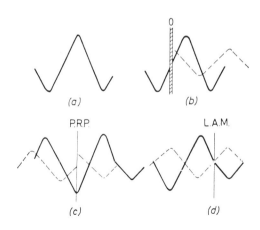

O. Object
P.R.P. Phase retarding plate
L.A.M. Light absorbing material

Figure 8.6 – Diagrammatic summary of the theory of phase-contrast microscopy

Although interference will now take place, the great difference in amplitude (or brightness) between the two sets of rays will prevent the maximum contrast from being obtained. To overcome this factor light-absorbing material is deposited in the area of the channel which reduces the amplitude of the direct light without affecting the diffracted light, thus permitting the maximum contrast to be obtained.

A broad summary of these principles is illustrated in *Figure 8.6;* (*a*) a ray of direct light from the annulus, on passing through the object (*b*) gives rise to a diffracted ray (dotted line), which is retarded by $\frac{1}{4}\lambda$; (*c*) on passing through the phase plate the diffracted ray, retarded by a further $\frac{1}{4}\lambda$, is now in a position to interfere with the direct light ray: (*d*) the amplitude of the direct light ray is reduced after passing through the light-absorbing material and better contrast is obtained. Although in the illustration, for the sake of clarification, (*c*) and (*d*) take place separately, in practice they occur almost simultaneously, and interference, of course, does not take place until the images are once again combined at the real image plane in the eyepiece.

The foregoing theory should only be true if monochromatic light is used as an illuminant, since white light would be split into its component colours when diffracted; in practice, however, white light may be used but better contrast is obtained by using a mercury green filter in conjunction with a compound high intensity lamp.

Equipment

Phase-contrast equipment need not be very expensive and several papers have been written describing methods of adapting normal microscopes for phase contrast at a small cost (Kempson, Thomas and Baker, 1948; Culling, 1950). The performance of such a converted microscope is not usually equal to a specially designed commercial model, but will suffice for routine purposes. The microscope described in this chapter will be that available from Vickers Ltd. *(Figure 8.7)*.

There are several other models available commercially which give equally good results by using a cross-shaped or slit-shaped aperture instead of an annulus in the substage, but since the underlying principle is the same they are not described.

Lamp

An intense source of illumination should be used, such as a high intensity compound lamp, with a mercury green (Wratten 62) filter.

Annulus

A different sized annulus will be required for each objective. These may be inserted separately, but a rotary changer carrying a set of annuli,

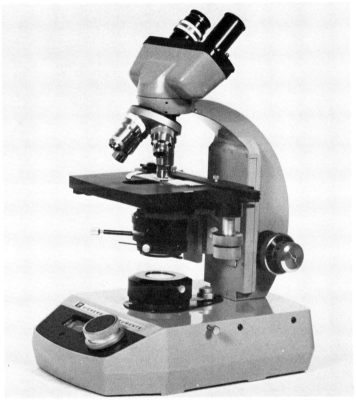

Figure 8.7 – The phase-contrast microscope (reproduced by courtesy of Vickers Limited)

one for each objective, mounted below a special condenser is more convenient each of these annuli can be centred by means of two centring screws.

Objectives

Objectives are supplied with phase plates already fitted, and since the phase plates affect their performance only slightly when used in a

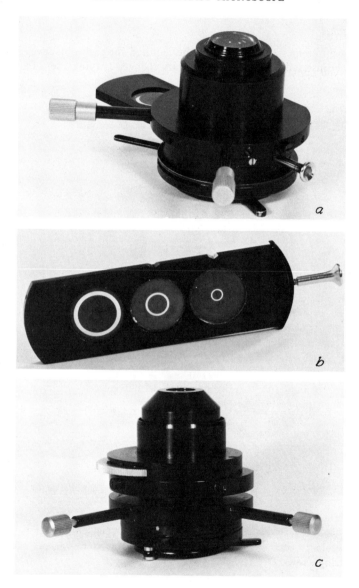

Figure 8.8 — The phase-contrast microscope; (a) and (b) show one way of inserting the phase rings into the light path, whilst (c) shows an alternative, possibly more popular way

normal manner, they may be used without an annulus for routine microscopy.

Auxiliary Microscope

The auxiliary microscope is used for examining the back focal plane of the objective and ensuring that the objective phase plate and the condenser annulus are properly adjusted.

Setting up the Microscope

(1)   The microscope is set up in the usual way (page 24), ensuring that there is no annulus in the substage.

(2)   Focus on the object, closing the iris diaphragm if necessary.

(3)   Rotate the annulus changer until the appropriate annulus is in position. Without disturbing the focus remove the eyepiece and replace it with the auxiliary microscope.

(4)   Adjust the auxiliary microscope to bring the image of the phase plate into sharp focus.

(5)   If the image of the light annulus does not coincide with the grey ring of the phase plate *(Figure 8.8a)*, it is adjusted with the centring screws until its image is concentric with, and completely covered by, the grey ring of the phase plate; the condenser may need to be raised or lowered slightly to adjust the size of the image of the light annulus. If the light annulus is not evenly illuminated the light should be adjusted.

(6)   The auxiliary microscope is replaced by the eyepiece and a phase contrast image will be observed.

Note. – This procedure should be repeated each time the objective is changed.

Chapter 9

The Nomarski Interference - Contrast Microscope

Phase-contrast microscopy has relatively recently (Nomarski, 1952, 1955) been supplemented by another technique for the microscopic examination of unstained biological materials, which has also been recommended as an adjunct to histochemistry (David and Williamson, 1971).

This method of visualization produces an image based upon optical path differences between λ/10 and a full wavelength, and therefore occupies a position midway between phase-contrast, which visualizes small path differences, and the conventional bright field microscope. By comparison to phase-contrast it presents an unconventional image which appears in relief. It must be remembered that this effect is optical and does not represent a deometric relief; for this reason the image should be interpreted with care. The diffraction halos seen with phase-contrast are not obvious with the Nomarski phase microscope, probably because when they are present they are dark rather than light.

The image may be viewed in light and dark contrast (using mono-chromatic light) or in interference colours using white light; in either case, optimum relief or satisfactory colour contrast can be obtained by adjustment of the position of the per prism.

Those readers who are interested in the detailed optical theory in-volved are recommended to read the paper by Padawar (1967) or the series of papers by Water Lang which are available from Carl Zeiss of Oberkochen, West Germany. Briefly, the method is based upon the interference of two mutually perpendicular plane polarized light waves. In the Nomarski phase microscope, the splitting of the light beam is achieved by the use of a modified Wollaston prism and not a bire-fringent plate as in the interference microscope (page 103). With a birefringent plate the beam splitting is lateral *(Figure 9.1)* whereas in the Wollaston prism it is angular *(Figure 9.1a)*. Furthermore, this deflection is at a relatively small angle, such that the distance between the two rays is very small and therefore both beams travel through the specimen and there is no 'reference beam' as with the Baker interference microscope (*see* page 105).

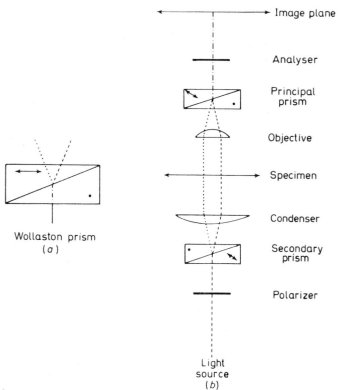

Figure 9.1 — (a) Conventional Wollaston prism. (b) Simplified diagram of the optical path of the Normarski interference-contrast microscope

The Wollaston Prism

The Wollaston prism consists of two prisms of a uni-axial, birefringent material (calcite or quartz) cemented together; the optical axes of the two prisms are at right-angles to each other. Using conventional Wollaston prisms only low-power objectives could be used; however,

Normarski modified these prisms so that the optical axis was inclined at a certain angle (lower component of secondary prism in *Figure 9.1b* is shown at an angle of 45 degrees instead of parallel to the upper surface as it would be in a conventional Wollaston prism *Figure 9.1a*). This modification, because of the position of its interference plane, allows the use of high-power objectives. As will be seen in *Figure 9.1b*, light rays are plane polarized by the polarizing filter (in the substage), split into two components (polarized at right-angles to each other) by the lower (secondary) Nomarski prism, and after passing through the specimen (which will give rise to path differences) they are then recombined by the upper (principal) Nomarski prism (by shifting this prism out of its centre position, an additional path differences can be superimposed upon the object path difference), the analyser then prepares them to interfere in the image plane.

Equipment

The equipment required *(Figure 9.2)* comprises a polarizer, the phase-contrast interference-contrast condenser (containing the secondary prism) with N.1. 1.4 front lens and an interference-contrast slide (containing the principal Nomarski prism and also the analyser above it). The prism must be appropriate for the microscope (or intermediate tube being used) to ensure that the interference plane of the Nomarski

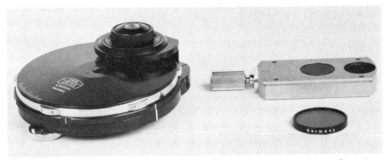

Figure 9.2 – Equipment for Nomarski interference contrast microscopy (reproduced by courtesy of Carl Zeiss, Oberkochen, West Germany)

prism lies in the focal plane of the objective. While a single principal prism can be used for X16, X40 and X100 objectives, a different secondary (condenser) prism must be used for each objective. Since a routine Zeiss microscope does not contain a slot for the interference slide an intermediate tube (with such a slot) may be inserted between the binocular attachment and the body of the microscope.

THE NOMARSKI INTERFERENCE-CONTRAST MICROSCOPE

Setting up the Microscope

(1) *Insert the polarizing filter* in the substage filter carrier, between the light source and the condenser. The vibration direction must be from right to left, or east-west with the microscope operator facing north.

(2) *Insert the interference-contrast (Inco) slide* in the slot above the objective, ensure that it is in as far as it will go and tighten the screw to fix the slide in position.

(3) Remove the condenser, auxiliary lens, objective and eyepiece. Looking down the body-tube, carefully *rotate the polarizer until the interference fringes are imaged with optimum sharpness in the field. If the dark fringe does not appear in the centre of the field, centre it with the control screw.* (These fringes are seen because we are viewing the modified Wollaston prism between crossed polarizer and analyser. The fringe system is parallel to the edges of the wedge with the central fringe being dark, while on both sides the fringes are of equally high order. The fringes result from the path difference between the two sets of waves from the prism.)

(4) Insert the interference condenser, objective and eyepieces. Check that the engraved arrows on the condenser and Inco slide point in the same direction. *Pull the Inco slide so that the prism is out of the optical axis of the microscope.*

(5) *Set the condenser to position 1 and check the bright field image* with X16 objective. *Connect the N.1. 1.4 condenser front lens to the underside of the specimen slide with immersion oil.*
(6)

(6) *Push in the Inco slide,* so that it is operative, and *adjust the control screw until the object shows optimum relief or colour contrast is obtained.* Areas of greater optical density may be imaged as hills or valleys depending upon the observer.

(7) Change condenser setting when changing objectives. Position I for X16, II for X40 and III for X100 objectives. The lamp and condenser diaphragms are adjusted accordingly.

Chapter 10

The Interference
Microscope

Although a form of interference microscope has been in use for some years, it is only since World War II that it has been developed for use in the biological laboratory with increased sensitivity. It will now detect and accurately measure phase changes in an object of as little as $^1/_{300}$ of a wavelength.

There are two types currently in use in Great Britain; the Baker interference microscope (designed by Mr. F. H. Smith), and the Dyson interference microscope (available from Cooke Troughton and Simms Ltd.). Since these microscopes are based on similar principles, only the Baker model *(Figure 10.1)* will be discussed.

The basic difference between the interference microscope and the phase-contrast microscope is that the former does not rely on diffraction by the object for interference, but generates mutually interfering beams which produce the contrast. It is this feature which enables such small phase changes to be seen and measured. The two rays, which eventually combine to produce the final image, are formed by a plate of birefringent material immediately above the condenser. These two rays, having passed through the condenser, are re-combined by a similar plate of birefringent material at the face of the objective *(Figure 10.2)*. Both these rays will have arisen from the same point of the light source, which is essential if interference is to take place in the final image, but the first will pass through one point in the image and the other through an area adjacent to it (the reference or comparison area). Consequently, each point in the final image is a compound one made up of two

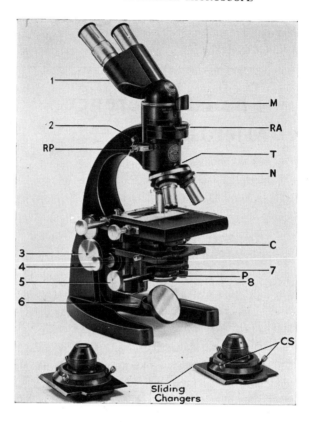

1. Binocular eyepiece.
2. Body-tube lock.
3. Coarse adjustment.
4. Fine adjustment.
5. Substage focus.
6. Mirror bracket.

7. Substage iris.
8. Filter ring.
RP. Quarter-wave plate.
M. Scale magnifier.
RA. Rotating analyser.
T. Tubelength corrector (on left.)

N. Nosepiece changer.
C. Condenser.
P. Polarizer.
CS. Condenser-adjusting screws.

Figure 10.1 — Double-refracting interference microscope for transmitted illumination

superimposed mutually different views of the same point of the object. By using polarized light and an analyser the phase relationship between the two rays can be adjusted and measured. The goniometer analyser is calibrated in degrees, and a magnifier fixed above enables small movements to be read accurately.

If white light is used as an illuminant, the various phases appear as different colours which change as the goniometer is rotated. Monochromatic light produces an amplitude contrast image and by rotation of the goniometer the degree of contrast can be varied for different components.

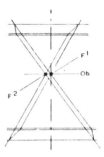

Figure 10.2 – Shearing system: Ob–object plane; F¹–object beam focus; F²–reference beam focus

Uses

The microscope can be used for two purposes. (1) As an infinitely variable phase-contrast microscope with which individual parts of living cells may be studied with maximum detail; for this purpose there is little to be gained over the use of a good phase-contrast microscope, particularly since the condensers and objectives are matched and a change of objective also means changing and adjusting the condenser. A further disadvantage is that the high power ($\times$ 100) objective is a water immersion lens with the consequent limitation of resolution, and so on.

(2) As a highly accurate optical balance, it may be used for estimating dry mass down to 1×10^{-14} g. The discovery that an increase in refractive index of 0.0018 is due to a 1 per cent increase in concentration of the solid substances contained in cells, and the fact that the refractive index of cell components can be estimated from the phase difference between them and the reference area (usually the fluid in which they are suspended) has made this possible. It is this aspect of interference microscopy that is at present being developed by many research workers.

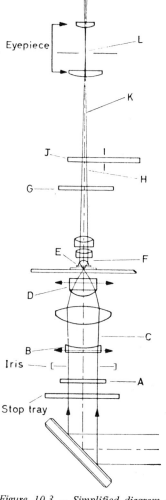

Eyepiece

L

K

J

I

H

G

E

F

D

C

B

Iris

A

Stop tray

A. Swing-out polarizer. The rotation of this element controls the intensity relationship between the double-refracted beams, permitting the out-of-focus image to be extinguished for normal transmitted light conditions.
B. Double-refracting negative lens.
C. The double-refracted rays entering the Abbé condenser.
D. Double-refracting plate cemented to the front lens of the condenser, rendering it bi-focal.
E. Double-refracting plate rendering the objective bi-focal.
F. The re-combined double-refracted rays.
G. Quarter-wave plate.
H. The re-combined rays circularly polarized in opposite directions by the quarter-wave plate.
J. Rotatable analyser, with swing-out section, calibrated in degrees.
K. The phase relationship between the circularly polarized rays is adjusted by the analyser.
L. Final image exhibiting interference between the in-focus image of the object superimposed upon the out-of-focus image.

Figure 10.3 — Simplified diagram of the double refraction interference microscope for X 100 double-focus system

(Figures 10.1, 10.2 and 10.3 are reproduced by Courtesy of Messrs. C. Baker, Ltd.)

Construction

Figure 10.3 is a diagram of the component parts of the Baker interference microscope and it will be seen that the instrument is an extremely complicated one, the use and adjustment of which requires a great deal of practice.

Chapter 11

Photomicrography

Microphotography is the process of taking minute photographs. Photomicrography is the process of taking photographs through a microscope, and it is surprising how often the former term is used to described the latter process.

APPARATUS

The apparatus used for photomicrography must be free from vibrations (or it should vibrate as a whole) and the source of light, condensing system, microscope and camera must be capable of being centred on the same axis, with the plane of the photograpidc plate (or film) parallel to that of the microscope stage. Unless the arrangement is temporary, it should be possible to fix all of these elements accurately in this position, although the camera should be removable to allow visual inspection through the microscope. With such an arrangement, it is preferable to have no physical contact between the microscope and camera, but to use a matte black light-tight connector between them. If the stand is vertical, a tube of black cloth can easily be made and is usually satisfactory. Elaborate apparatus is not a necessity and very good work can be done with simple equipment. *The microscope must, whatever method or apparatus is used, be critically illuminated using Köhler illumination* (*see* page 24).

Elimination of Vibration

The ideal location for a photomicrographic outfit is, of course, in a place where there are no vibrations from the outside and which is itself directly part of a massive support. If this is not possible, the camera support must be stable and preferably supported separately.

107

If persistent trouble from outside vibration is encountered, it is wise to mount the whole photomicrographic table or bench on a suitable absorbent for the vibration. This is better than mounting the microscope or even the microscope and the camera on such an absorbent, since the floating system should have as great an inertia as is conveniently possible. The best type of vibration absorbent varies with the source and type of the incoming vibration. Sponge rubber is often used and is helpful if it is sufficiently thick and is renewed frequently; bath sponges are said to represent the most effective type even if they are apparently flattened beyond usefulness by the weight. Horizontal vibrations seem to be best removed by standing the table on multiple layers of some kind of fabric or metal sheeting.

Apparatus for Simplified Photomicrography

A relatively inexpensive yet convenient set-up can be made from a wooden board and the usual type of metal laboratory supports, although the vertical support rod must be of adequate diameter; that of the usual $3/8$ or $1/2$-inch laboratory rod is insufficient. The camera can be supported on its face by a horizontal wooden board which has one or more holes cut through it to allow the camera bellows or other projections to drop through. The board in turn is supported by a laboratory extension ring (such as is frequently used with flasks) which swivels around the vertical support stand and thus can be swung aside during the usual use of the microscope. The camera can be clamped into position by heavy rubber bands snapped over picture hooks.

The greatest simplification of all consists in the use of a roll-film camera since with it a darkroom is not even necessary. Not only is the camera loaded in daylight, but the film can be sent to a photofinisher for development and printing. A personal camera can be used with photomicrographs and general pictures on the same roll of film.

Use of Camera with Lens

A compound microscope is designed, in general, for visual work with its image at infinity, that is, with parallel bundles of rays emerging from the eyepiece. The eye is expected to focus this image on the retina while the eye is relaxed, as it is when it is focused on distant objects. Therefore, if the camera with its lens focused at infinity is placed over a well-set-up microscope that has been visually focused, a reasonable photographic reproduction of the image can be obtained by exposure of the film. In practice it has been found, however, that most people when looking through a microscope strain a trifle with their eyes; therefore, it is best to focus the camera at 25 feet.

One further precaution should be taken. The distance of the camera should be so adjusted that the eyepoint of the microscope falls in the centre of the front surface of the camera lens. This point can be determined by placing a thin sheet of paper above the eyepiece and moving it up and down until the circle of light is the smallest. The definition is best with this arrangement, and the quality of the lens is least important in this case. The magnification will be less than the rated visual magnification of the compound microscope. With a lens of average focal length in a hand camera, the magnification will be about one-third of its rated value.

Sometimes it takes a picture of the illuminated back lens of the objective and superimposes that as a small flare spot on the picture. Raising the camera from the plane of the eyepoint will cause this flare image to go out of focus rapidly; the field then becomes smaller on the film as it is limited by the lens diaphragm, and the definition becomes somewhat degraded.

In the absence of proper equipment, the convenience of the method often outweighs these disadvantages. The use of a single lens reflex camera will usually obviate most of these difficulties and give reasonable results.

Use of a Camera Without a Lens

A camera is most frequently used in this manner for photomicrography, either with a special attachment *(see below)* or by focusing the image on a ground glass in the same plane as the photographic emulsion. The latter is achieved by removal of the ground glass and its replacement by the camera, most conveniently on a metal slide. For accurate focusing by this method it is better to focus the image with a magnifying lens against a clear portion of the ground glass bearing cross lines. This can be made easily by cementing a micro-coverslip to the ground glass side with Canada balsam, after having first drawn a well-centred cross on the ground glass with a pencil.

When it is not desirable to have a permanent clear area in the ground glass of a camera, the coverslip can be held with cedarwood oil. The film plane of such a camera should be set up no nearer than 10 inches from the ocular (except as discussed below). It should be rigidly supported so that the extended optic axis of the microscope intersects the cross line on the ground glass.

Photomicrographic Attachments

While the automatic apparatus is ideal, good photomicrographs may be obtained with a special camera attachment to the routine microscope; the only limitations being those imposed by the lens system being used (achromatic, aplanatic, and so on) and the operator. Such an attachment should have a camera body (to carry the film) and attach directly into the body tube of the microscope. A monocular, or tri-ocular attachment should be used; one of the tubes of a binocular can be used (with the other one covered) but will require a longer exposure. The attachment usually consists of a direct tube which attaches to the camera body; the other end, usually containing a X10 eyepiece fits into the microscope body tube. The direct tube should have an observation side-arm fitted with an adjustable (focusing) eyepiece. On looking through the observation eyepiece, cross-hairs or a pattern of double lines will be seen, these should be in sharp focus at the same time as the object to be photographed. The milled ring on the side-arm eyepiece is rotated until the lines are sharp; thus, when the object (viewed through the side-arm eyepiece) is focused with the microscope fine adjustment, it will be in focus on the film emulsion. Most photographers prefer to use an attachment with a special, non-vibrating camera shutter such as the Compur for photomicrography. In use one opens the camera shutter using the time (T) or bulb (B) mechanism, makes the correct exposures with the Compur shutter, then closes the camera shutter and advances the film ready for the next exposure.

Automatic Photomicrography

In 1955, Carl Zeiss introduced what was probably the first completely automatic photomicroscope. Our Department has had one of these models since 1958 and surprisingly for what seemed to be such a complex piece of apparatus, it has been virtually trouble-free over the intervening years. During this period it has often been used to give good results by almost completely inexperienced microscopists. Almost every microscope manufacturer now produces a form of automatic photomicroscope, either as an attachment for a research microscope or as a complete unit.

The modern edition of this type of apparatus *(Figure 11.1)*: (1) uses a photomultiplier instead of a photocell (which gives more accurate readings); (2) allows the exposure to be based either upon the overall lighting of the specimen field or upon a particular area of the field; (3) allows for a greater variation in the speed of the film being used (with

ASA ranges from 5 to several thousand) and; (4) has shutter systems and film transport mechanisms which have been markedly improved. Most new machines will allow deliberate over-exposure or under-exposure if desired, which was not the case with the earlier models.

No attempt will be made to describe any of the available models in detail since they all vary, to a greater or lesser degree, in operation and they are almost certainly available with the type of microscope preferred by the reader. The particular company concerned will willingly supply descriptive pamphlets and arrange for a demonstration of the apparatus.

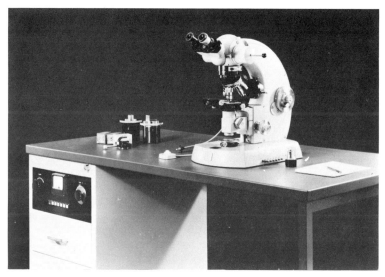

Figure 11.1 — Zeiss Photomicroscope II, combination of research microscope and miniature camera (24 × 36 mm) for automatic exposure control from 1/100 second (reproduced by courtesy of Carl Zeiss, Oberkochen, West Germany)

The great advantage of these microscopes is that they can be used as a routine microscopes, and any unusual or interesting specimens may be photographed by the touch of a button. By using two film cassettes either black and white or colour photographs can be taken by simply changing cassettes and making a minor adjustment to the controls. It will follow that it is essential to correctly illuminate and align the micro-scope optical system (*see* page 24) *and most important of all to focus the eyepieces on the cross hairs* (or similar markings) to ensure that the image will be in focus at the plane of the film; most of the problems of badly focused photographs are due to having forgotten this elementary precaution.

LIGHT SOURCES

Most of the better illuminants sold for visual work with the microscope can be used, particularly for low-power work in a bright field. If considerable photomicrographic work is to be done, however, one of the lamps especially made for this purpose should be obtained. The more frequent use of narrow-band filters in photomicrogrphy is one reason for the need of a more intense light source of photomicrographic than for visual work.

For photomicrography with oblique, reflected illumination the problem is usually to get the beams of light at the objects with satisfactory control of direction and quantity because of limited working space. Special illumination devices for this type of work are now commercially available from several microscope companies.

Such devices are enormously time saving and convenient compared with the set-up and manipulation of the ordinary type of lamps. For low-power work, they usually consist of either a ring reflector or a circle of illumination surrounding the objective. Such a device can be constructed with flashlight or automobile headlight bulbs. For medium and high powers, special units consisting of an objective surrounded by a ring condenser are employed, the illumination from a lamp at one side being utilized by a proper optical system. These are not only convenient, but at higher magnifications where there is little working distance, it is impossible to obtain comparable results by other expedients. Directions for the operation of these devices can be obtained from their manufacturers.

Whenever a very high intensity light source is used, such as a ribbon filament lamp, it is a wise precaution to place a heat-absorbing filter in the light beam. This is to protect the specimens, the cemented objectives and the thick lenses of the modern high-aperture substage condenser from the intense concentration of heat. A 2-inch layer of water is usually sufficient, but filling the cell with a 1 per cent solution of copper sulphate acidified with a drop of sulphuric acid is more effective, although achieved at the cost of a little loss in a white light exposure with panchromatic film. A piece of special heat-absorbing glass can also be used. Coloured heat filters cannot, of course, be used for colour photography.

Exposure Times

A certain amount of latitude in exposure time is possible with modern photographic materials. As will be seen on page 122; however, fine structure can be lost if a negative is badly over-exposed or under-exposed. A properly exposed negative will show excellent gradation from almost black to almost clear, should this gradation be absent the specimen should be re-photographed.

Trial Exposures

If the equipment used is standardized, for example, the same microscope, lenses, lamp setting (with variable control), condenser and lamp diaphragms, camera attachment and film type (and speed), then by a series of trial exposures the approximate exposure time for each of the objectives can be determined.

The method, which is simple and logical, is to take a series of doubling exposures with each objective, noting carefully whether top lens of condenser is in or out, degree of diaphragm openings — *these must be adhered to subsequently.* Such a series with a low power (×10) objective would be (in seconds) 1, 1/2, 1/5, 1/10, 1/20, 1/50, 1/100, 1/250, the actual times used would depend upon the shutter speeds available on the equipment being used. The specimen photographed should be typical of the type likely to be encountered in the reader's particular field with a reasonable gradation of contrast. The test film is then developed and after examination and printing of the negatives a satisfactory exposure determined for each objective. This information should be typed upon a card which is kept with the microscope. While this method will work with colour transparencies — the emulsion is not as tolerant as black and white film — it is better to take two different exposures for each field photographed. In one laboratory I used this method for almost 2 years with reasonable success.

Photoelectric Exposure Meters

There are now available several excellent exposure meters designed for the low intensity levels of microscopic illumination. Microscope manufacturers will either supply, or recommend, models for use with their equipment. It should be noted carefully where the reading is to be taken, for example, directly from the microscope body tube, through the normal eyepiece, or the eyepiece of the observation side arm, as the light intensity may vary greatly between them.

Fluorescent photomicrography presents special problems since the light intensity may be so low that few exposure meters will read it accurately, a further difficulty being that quite often the extended exposures required 'burn out' the fluorescence in the specimen. It is best to use the most rapid film available in order to keep the exposure as short as possible.

MICROSCOPE LENSES IN PHOTOMICROGRAPHY

Objectives (*see* page 16)

Routine achromatic objectives may be used, especially if monochromatic light is used; for example, green filter for photographing haematoxylin and eosin stained preparations. Such objectives will give a sharp picture in the centre of the field with some loss at the periphery whereas it is possible to purchase *planachromatic objectives* which are designed to give a flat field for photomicrography. These are especially good for black and white photography, but for colour work the *planapochromats* are to be preferred because of their better colour correction.

Oculars (*see* page 9)

While normal oculars may be used in properly designed photomicrographic equipment, a projection ocular should be used if a bellows extension is fitted. This type of ocular has an adjustable top lens which is used to focus the image in the ground glass (at the top of the bellows) without changing the tube length of the microscope. One focuses the object with one eye in the usual manner, then adjusts the top lens to focus the image on the plate glass.

Magnification

To calculate the magnification obtained on a photomicrography negative it is necessary to know (or measure) the distance (in inches) from the ocular to the film plane in the camera; the magnification of the objective X eyepiece is then divided by 10 and multiplied by the distance between the ocular and the film; for example, with a X10 objective and X10 eyepiece and the film 5 inches from the top lens of the eyepiece the magnification will be $10 \times 10 \times 10/5 = 50$. This calculation is necessary because the eye forms a virtual image at a distance of 10 inches (*see* page 28) of the real image formed in the ocular, and to obtain a similar real image it is necessary to place a ground glass (or film) at a distance of 10 inches from the eyepiece; a ground glass or film at a distance of only 5 inches from the ocular will receive an image only one-half of the size seen at a distance of 10 inches.

Photomicrographic Detail

Detail in a contact print may be finer than the eye can see and as such, be wasted; conversely, it may need to be enlarged beyond the useful limits of magnification because the numerical aperture of the objective used was too low to produce the resolution (detail) required (*see* page 13). The unaided human eye can normally resolve (or separate) distances of 100–250 microns; lines or dots closer together than this distance will appear as a continuous structure and even with 250 micron separation good contrast (for example, black on white) is required for visual resolution. The formula for computing the maximum useful magnification possible for a given numerical aperture (NA) or the NA needed for a given magnification (Mag); where Wl = wavelength of light used (usually 0.5 microns) and D = diameter of antipoint (or separation distance with unaided eye, usually 250 microns) is:

$$NA = \frac{Wl \times Mag}{2 D} = \frac{0.5 \times Mag}{500} = \frac{Mag}{1,000}$$

Therefore, to determine the NA needed for a given magnification, divide the Mag. by 1,000; to determine the magnification possible for a given NA, multiply the NA by 1,000.

Low-power Photomicrography

Low-power pictures may be taken either by (1) the addition of close-up lenses to a normal camera; or (2) by the use of low power objectives.

The addition of close-up lenses to a normal camera gives greater depth of focus and a larger field of view than (2) but with less definition because of the relatively poor NA of the camera lens, for example, f/6.3 = NA/0.08; f/4.5 = NA of 0.11; f/1.6 = NA of 0.31. As a general rule if there is little detail in the specimen (and depth of focus is needed) a camera lens should be used; if there is a large amount of detail (and depth of focus is not important) a low-power objective should be used; if necessary, stop down objective to give greater depth

115

of focus although this will reduce the NA. The highest useful magnification with f/4.5 is ✕35, and with f/2.5 is ✕50.

IMAGE CONTRAST

One of the most important factors in photomicrography is the contrast – or the brightness differences – in the image and also in the corresponding tones of the photograph. The contrast within the different specimens varies greatly, as does the contrast between the structure to be shown and its background. Therefore, the problem usually resolves itself into the rendering of intensified or diminished contrast. The contrast of the negative depends upon both image contrast and the photographic contrast of the plate. In photomicrography, both of these factors are, to a large extent, under the control of the operator. The control of contrast in the formation of the microscopic image is now discussed, the control of contrast in the photographic image will be dealt with separately.

Causes Preventing Sufficient Contrast in the Image

Dirty lenses or cover glass will spoil the whole image, covering it with scattered light ruinous to all contrast. Sometimes insufficient contrast may be produced by unsatisfactory lighting; too much scattered light from illuminated portions of the specimen outside of the field of view, or a larger condenser stop than the objective will bear, will produce it. A field diaphragm will obviate the first difficulty. In bright field work the sub-stage iris diaphragm should never be opened more than is necessary for that amount of light to just fill the aperture of the objective in use. If it is necessary to control the visual contrast of colourless objects by the condenser diaphragm, remember that as the diaphragm is closed to gain image contrast, it is obtained at the expense of resolving -power. A superior method to control image contrast, applicable to all coloured specimens, will be found in the control, by filters, of the colour of the light used for illumination.

USE OF COLOUR FILTERS

Definition of Colour

Light is known to consist of waves, and the colour of the light is connected with the length of the waves. The length of a light wave is the distance from the crest of one wave to the crest of the next, measured in very small units. (The millicron (mμ) is one-thousandth of a micron (μ) and one-millionth of a millimetre (mm), which in turn is about one-twenty-fifth of an inch; an Angstrom unit (Å) is one-ten-thousandth of a micron.)

A wave of the darkest violet that we can see will be; 400 mμ in length; a wave of blue-green, 500 mμ; of bright green, 550 mμ; of orange, 600 mμ; and of deep red, 700 mμ. Visible light, then, is composed of light waves varying in length from 400 to 700 millimicrons which may be divided roughly into three portions: blue-violet, 400–500 mμ; green, 500–600 mμ; red, 600–700 mμ *(Figure 11.2)*.

The position of an absorption band may be defined by the length of the waves of light which are absorbed by it. We may speak of an absorption band as extending, for instance, from 600 to 640 units meaning thereby a band absorbing those particular waves of red light, and thereby producing a blue colour.

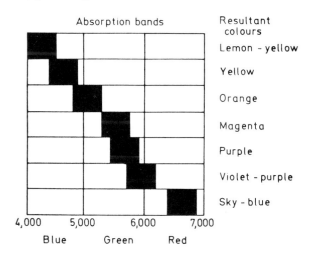

Figure 11.2 – *Colours resulting from absorption in specific areas*

Colour Produced by a Single Absorption Band

In order to appreciate the relationship between absorption and the resultant colour of the object, it may be worthwhile to examine the effects of a single sharp absorption band in different parts of the spectrum. First consider a sharp absorption band in the extreme red, stretching from 640 to 690 mμ, and completely stopping all light of those wavelengths. The remaining colour will consist of all the blue-violet and all the green light with some of the red. The actual visual effect of such a mixture, what may be called the residual colour, is a sky-blue. Imagine this band moving to where the absorption is between 580 and 620 mμ; the residual colour will be a light blue-violet because there is a great deal of red being transmitted and less green *(Figure 11.2)*.

Therefore, when anything is coloured sky-blue, it means that it is absorbing the deep red; a violet-blue object absorbs the orange, a purple the yellow-green, a magenta the central green, an orange the blue, a yellow the blue-violet.

If a sky-blue object is looked at through a piece of yellow glass, it will be found to look bright green in colour, therefore a green colour is produced by the absorption both of the red and of the blue, the blue object absorbing the red light and the yellow glass the blue light. If a colour is to be rendered as black as possible, then it must be viewed or photographed by light which is completely absorbed by the colour, that is, by light of the wavelengths comprised within its absorption band.

Contrast Within the Specimen Itself

The second rule for procedure deals with the case where contrast is required, not against the background but within the object itself.

A good example of this is the photography of an unstained section of whalebone; this is of a yellow colour and shows ample detail to the eye but it completely absorbs blue-violet light. If it is photographed only in the blue-violet light to which an ordinary plate is sensitive, then it shows far too much contrast, appearing as a black mass without detail against the background and presenting a exaggerated example of the loss of detail.

The proper procedure in this case would be to photograph the object by the light which it transmits. A whalebone section, for instance, photographed by red light will give perfectly satisfactory results and show ample detail in structure.

A class of microscopic objects which frequently requires such treatment are the insect preparations, which usually give the most satisfactory results when photographed by yellow or red light.

Procedure for Choosing Colour Filters

The best method for determining the degree of contrast required to photograph a particular object is to examine it visually under the microscope, first by means of a combination of filters transmitting as completely as possible light of the wavelength absorbed by the preparation, and then by other filters transmitting light less completely absorbed, until the degree of contrast obtained is satisfactory to the eye. As a guide to procedure, a list of stain colours and of the filters which will produce the maximum contrast with them is given below.

For	blue	stained	preparations	use	a	red	filter
··	green	··	··	··	··	red	··
··	red	··	··	··	··	green	··
··	yellow	··	··	··	··	blue	··
··	brown	··	··	··	··	blue	··
··	purple	··	··	··	··	green	··
··	violet	··	··	··	··	yellow	··

When selecting a filter, care should be taken to avoid excessive contrast as this will result in a loss of detail.

SELECTING AND DEVELOPING FILMS

The science of sensitometry is concerned with the description and measurement of the photographic properties of sensitized materials, and to read photographic data sheets some knowledge of the methods and the language employed will prove useful.

The Characteristic Curve

The standard method of presenting sensitometric data is with the characteristic curve. This is a plot of the response, optical density, versus the logarithm of the exposure. Such a curve is obtained by subjecting the photographic material to a series of exposures, each greater by a constant factor than the preceding exposure (for example, $\times 2$), and then processing the material and reading the resultant densities with a densitometer *(Figure 11.3)*. When the density of each silver deposit is plotted against the logarithm of the exposure which produced that density, a curve can be drawn through the points so plotted. This curve is the 'characteristic curve'.

The portion of the curve from B to C is the range wherein the gradient is constant and the density increases as a direct or linear function of the logarithm of exposure. For the best photographic image, it is preferable that the exposure be placed on the straight-line portion of the characteristic curve (B–C). Also, to reduce graininess, the exposure range should be held as low on the straight-line portion as possible.

Gamma

The slope of the straight-line portion of the characteristic curve is designated as 'gamma' (γ). The numerical value of gamma is defined as the tangent of the angle made with the exposure axis *(Figure 11.3)*.

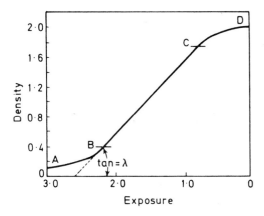

Figure 11.3 – The characteristic curve

Gamma serves as a convenient method of expressing contrast and depends partly on the characteristics of the emulsion and partly on the degree of development.

If characteristic curves are determined for a series of development times and the gamma of each curve is plotted against the time of development, a curve showing the rate of growth of gamma is obtained. With increasing development, a maximum value of gamma is obtained; over-development may produce a fall in gamma with an accompanying increase in fog density.

Reciprocity Effect

The reciprocity law, formulated by Bunsen and Roscoe without specific reference to photography, states that the product of a photochemical reaction is dependent on the total energy employed. This is to say that, in photography:

Exposure (E) = intensity of exposure (I) × time of exposure (t) and the two factors, I and t, act independently. This is generally true for photographic materials except for exposures to very low and very high levels of intensity (I). Most photographic materials show something of a loss in sensitivity when exposed to very low and very high levels. This loss in sensitivity is known as the 'reciprocity effect' or 'failure of the reciprocity law'. Since much scientific photographic work is done at very low or very high intensities, the reciprocity effect may become significant. It is therefore an important factor in the choice of materials, especially for use in photographic photometry.

Clayden Effect

If a photographic emulsion is given first a very short exposure to light of very high intensity and then a second exposure to light of moderate intensity, the two do not add in a simple fashion.

The very high-intensity exposure densensitizes the emulsion effectively toward the second exposure. If the first exposure has affected only a part of the emulsion, and the second exposure is uniform over all the emulsion, the image of the first often appears reversed upon development, that is, positive instead of a negative is formed. This phenomenon is known as the Clayden effect. It has been observed frequently in the photography of lightning flashes, where it gives rise to the so-called 'black lightning'.

Herschel Effect

If an emulsion that has not been dye-sensitized to red or infrared is exposed to blue or white light a latent image will be formed in the normal way. If the emulsion is subsequently exposed to red or infrared radiation before it is developed, some of the effect of the original exposure will be erased. Thus, the long wavelength radiation is capable of destroying, to some extent, the latent image formed by the blue light — this is the Herschel effect. It is not subject to reciprocity-law failure, so far as is known.

Resolving Power

The term 'resolving power' refers to the ability of an emulsion to record fine detail. In measuring resolving power, a parallel-line test chart is photographed at a greatly reduced scale. The lines of the test chart are separated by spaces of the same width as the lines. The image is examined under a microscope, and the number of dark lines per millimetre that are just recognizable is determined. Lines closer together (more lines per millimetre) that are indicated by this number will appear on the plate, not as individual lines but as an unresolved grey mass. The resolution of an emulsion depends only slightly on the degree of development.

Resolution falls off greatly at both high and low exposure levels reaching a maximum at some intermediate exposure; it is for this intermediate exposure that the resolving-power classification is given. The maximum is sharpest for emulsions of high resolving power. In general, resolving-power maxima occur in the density range from 0.7 to 1.6. The resolving powers of films and plates are described in data sheets. Measurements made on the same material in different laboratories may not agree, because of differences in equipment and procedure. Moreover, the characteristics of materials sometimes change slightly with manufacturing variations or as improvements are made. For these reasons, materials have not been given specific numerical values but have been placed in the following classifications.

Low resolving power includes films and plates with resolving powers below 56 lines per millimetre.

Moderately low resolving power includes materials capable of resolving from 56 to 68 lines per millimetre.

Medium resolving power includes materials capable of resolving from 69 to 95 lines per millimetre.

High resolving power includes materials having resolving powers between 96 and 135 lines per millimetre.

Very high resolving power includes materials with values between 136 and 225 lines per millimetre.

Extremely high resolving power. There are several special purpose materials, mostly of low speed and high contrast, having resolving powers above 225 lines per millimetre. They have been produced for special applications where resolution is of prime importance. The Kodak High Resolution Plate and the Kodak Spectroscopic Film Type 649–GH have resolving powers of 2,000 lines per millimetre or more.

Sharpness

Sharpness refers to the ability of photographic material to give a sharp boundary between areas receiving low and high exposures. It has been shown that, for photographs that are to be observed visually, sharpness is related to the mean of the square of the density gradients, D/X, across such a boundary in the developed image. The quantity resulting from this evaluation has been termed acutance. This theory is described by Higgins and Jones (1952, 1953).

Development

The recommended developer for general use with scientific and technical plates and films is Kodak Developer D–19, which gives very good contrast and high effective speed with low fogging tendency, Kodak Developer D–19 has very good keeping qualities and a high useful developing capacity, especially when used with Kodak Replenisher D–19R.

For some products which would require relatively long development times in Kodak Developer D–19, alternative recommendations are given for the use of Kodak Developer D–8, with much shorter development times. However, Kodak Developer D–8 has a much lower useful developing capacity and somewhat poorer keeping qualities than D–19. The use of D–8 usually results in a higher contrast image than is obtainable with D–19.

Agitation

The most frequent cause of processing variations, with consequent variation of results, is improper agitation techniques during development. Only by proper agitation can uniform development and freedom from flow marks be assured.

For a low volume of processing where high quality and accuracy are important, an effective method for processing films and plates with uniform results is to brush the emulsion surface, slowly and continuously during development, with a soft camel's hair brush of greater width than the plate.

Stop Baths

Since high energy developers are usually employed in scientific and technical work, and uniform development is necessary for precise work, it is recommended that the development be stopped completely and quickly by immersing films and plates in a stop bath (such as Kodak SB–5). Films and plates should be rinsed, with continuous agitation, for at least 30 seconds before transfer to the fixing bath. While a stop-bath rinse is preferred and plates can be rinsed with continuous agitation in running water at $65-70°F$ ($18-21°C$).

For general use, Kodak Fixer* or Kodak Fixing Bath F–5 are recommended. When more rapid fixing is desired, Kodak Rapid Fixer*

*Available in prepared packages.

or Kodak Rapid Fixing Bath F–7 can be used. Films and plates must be agitated frequently during fixing at a temperature of 65–70°F (18–21°C).

Washing

In high quality processing, proper washing is very important. Films and plates should be washed for at least 20–30 minutes in running water at 65–70°F (18–21°C). For more rapid washing, or where an image must be permanent for archival purposes, the use of a hypo clearing agent is recommended.

Drying should be done in a dust-free place. Any tendency for the formation of drying marks can be minimized by treating the films or plates in diluted Kodak Photo-Flo Solution after washing or by wiping the surfaces carefully with a damp photo chamois or soft viscose sponge.

Colour Film

A new 35 mm colour film for transparencies has recently been made available from Eastman Kodak; it is Kodak Colour Film SO–456. This is an extremely fine grain, slow speed (ASA 16), high definition colour reversal film. It has high contrast and good colour saturation characteristics. Its most striking qualities are the absence of grain, the clear intense colours and the almost white background (instead of the usual blue-grey). Details are available from Eastman Kodak Company, Rochester, N.Y. 14650, USA, or Kodak Canada, Limited, 3500 Eglinton Avenue West, Toronto 15, Ontario, Canada.

In addition to the above film which we have only recently obtained, we routinely use Ektachrome, and High Speed Ektachrome (Daylight) for fluorescence photomicrography.

Chapter 12

The Electron Microscope

Introduction

The light microscope lens systems have been perfected to such a degree that resolution is limited by the physical properties of image formation rather than the properties of the lenses themselves. From the discussion of optical theory it was seen that the resolution of any optical system is limited by the wavelength of light employed, and that an object that is smaller than this wavelength will cause so little perturbation of the light beam that it will not be resolved in the image. The best light microscopes are therefore limited to a resolution of about 2,000 Å*. The ultraviolet microscope, by using wavelengths about one-half that of white light, achieves a resolution of 1,000 Å.

By using an electron beam instead of light rays, the electron microscope gives much better resolution. The wavelength of moving electrons depends on their velocity. At an acceleration of 50,000 volts they have a wavelength of about 0.01 Å, and one may expect to resolve images of about this order. Due to lens defects which can be corrected in the light microscope but have not so far been corrected in the electron microscope, the resolution is limited to about 4 Å, which is still several orders of magnitude better than the best optical microscopes.

Tissue to be examined in the electron microscope must be processed so that sections of less than 0.1 μ thickness can be cut. These very thin sections are necessary because of the poor penetrating properties of the electron beam; the usual sections prepared for light microscopy would be completely opaque. Techniques of tissue preparation have now been developed so that it is relatively easy to cut thin enough sections, and

*1 mm = 1,000 microns; 1 micron = 10,000 Angstrom (Å).

basically these techniques are the same as those used in routine histological laboratories. Briefly they consist of fixation, dehydration in alcohol, embedding in special plastics, and sectioning with a specially designed microtome, using knives made from broken glass. The thin

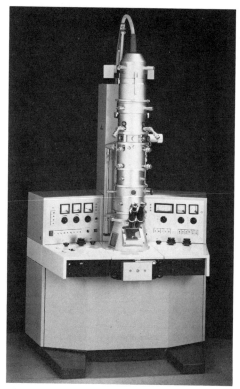

Figure 12.1 – Sieman's electron microscope

sections are picked up on small copper grids for examination in the electron microscope.

Before describing briefly these new techniques, some elementary principles of the electron microscope will be reviewed.

Theory and Construction of Electron Microscope

The convergence of a light beam by a convex glass lens has its counterpart in the convergence of an electron beam as it passes through the core of a circular magnetic field. Most electron microscopes use electromagnetic lenses. The convergence of an electron beam is shown producing an image; the image and object distances are related to the focal length of the lens in exactly the same way as in light optics (*see* page 3). The electron microscope is therefore constructed on the same optical principles as the light microscope, and the same formulae can be used to correlate magnification and focal distances *(Figure 12.2)*.

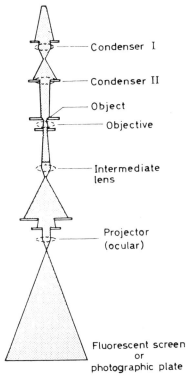

Figure 12.2 – Diagrammatic illustration of the optical path of the electron microscope

The electron beam is obtained from a heated tungsten filament which is surrounded by a metal cylinder known as the Whenalt cap. This cap serves to shape the electron beam. Just beyond the Whenalt cap is the anode which has an aperture through which the electron beam passes. A large voltage is applied between the cathode (the tungsten filament) and the anode, this gives the electrons their high velocity.

They pass through the rest of the microscope without any further acceleration. The Whenalt cap is given a voltage slightly lower than the filament, and this voltage is usually variable so that the flow of electrons from the cathode can be controlled. This is known as the bias voltage.

The electron beam first passes through the condenser lens. As in the light microscope this lens serves to focus the beam on to the object, and so provide 'illumination'. One must remember that the magnetic lens of an electron microscope can have different powers depending on the amount of current flowing in the electrical coils. In the light microscope the lens powers are, of course, fixed but the lenses are made movable with respect to the object so that the image can be focused and proper conditions of illumination obtained. In the electron microscope all of the lenses are rigidly fixed, but their focal points are variable by adjusting the lens currents. Thus, the 'illumination' of the object is achieved by varying the current in the condenser lens.

The imaging system of the electron microscope usually consists of three lenses; the objective, intermediate and projector lens. This gives three stages of magnification and makes it possible to achieve high magnification in a reasonable amount of space. The objective lens is placed with its focal point close to the object. Intermediate images are formed between each lens. The projector throws its image on to a fluorescent screen which may be substituted by a photographic plate to make a permanent record *(Figure 12.3)*. The entire illuminating and imaging system is usually referred to as the microscope column and is constructed upside down compared to the light microscope; that is, the electron gun and condenser lens are placed above and the image is formed below *(Figure 12.1 and 12.2)*. The column is very rigidly constructed and is maintained in a high vacuum since air molecules would deflect the electron beam. Because the specimen must be placed inside the vacuum it is not possible to examine living material in the electron microscope.

Electron optics are essentially similar to light optics. One important difference, however, is that the formation of the image is due to scattering of electrons by the molecules of the specimen and this scattering depends solely on the mass densities. Elements of high atomic weight such as lead or uranium cause marked electron scatter and appear very dense in the electron image. The lighter elements such as carbon, oxygen and nitrogen cause little electron scatter and have poor contrast. In the light microscope the image is due to absorption of light which depends more on molecular structure than atomic weights. Histological stains depend on absorption of certain wavelengths of light due to their molecular structure and are composed mainly of carbon, nitrogen and hydrogen atoms. Since these are all of low atomic weight they have little electron scattering power and are not generally useful as stains for electron microscopy. Unstained tissues have very poor contrast in the electron microscope, but may be stained by a variety of heavy metal salts. Most such 'electron stains' are relatively non-specific and one does not have the battery of stains which are so useful in studying tissues in the light microscope.

Figure 12.3 – Electron micrograph of re-precipitated collagen fibres from rat skin which have been shadowed with platinum. The white striations are approximately 2,400 Å apart

PREPARATION OF TISSUES FOR ELECTRON MICROSCOPY

The general principles are the same as those in the routine histological laboratory, and consist of fixation, dehydration, embedding, sectioning, and staining. The following pages will describe the methods which are in common use in most electron microscope laboratories. Many special and more or less experimental techniques are also available. For enzyme histochemistry the reader is referred to the papers by Holt and Hicks (1961), for antibody staining the paper of Singer and Schick (1961), for the preparation of frozen-dried tissues, the papers of Gersh (1956) and Grunbaum and Wellings (1960), and for techniques of freeze etching to the papers of Moor and his colleagues (1963, 1964).

Fixation

The most commonly used fixative is a buffered solution of osmium tetroxide originally described by Palade and often referred to as Palade's fixative. The preparation is given overleaf.

Palade's fixative

Stock buffer

Sodium barbitol	14.7 g
Sodium acetate	9.7 g
Distilled water to make	500 ml

Formula

Stock buffer	5 ml
0.1 N HCl	5 ml
Distilled water	2.5 ml
2 per cent aqueous osmium tetroxide	12.5 ml

The pH should be 7.4. The buffer solution should be kept in the refrigerator and is good for several months. The fixative should be used within a week or so and kept in a dark bottle in the refrigerator. *Osmium tetroxide fumes are extremely toxic. Avoid breathing the fumes or getting any solution on the hands.* The 2 per cent solution is best made by dropping a cleaned ampule into a glass-stoppered bottle containing the necessary quantity of water, and breaking the ampule by vigorous shaking. Solution of the crystals can be hastened by gentle heat.

Since Palade's fixative is hypotonic, many laboratories add sucrose, 0.045 g per ml. This is then known as Caulfield's fixative. Fixation in osmium tetroxide should not be longer than 60 minutes.

Glutaraldehyde Fixation

Glutaraldehyde fixation was introduced by Sabatini and colleagues (1963) as a method for electron microscopy. It is probably one of the most commonly used fixatives in this field at the present time, being customarily employed together with post-fixation in osmium tetroxide. Dr. W. H. Chase, in our laboratory, routinely fixes tissues in buffered glutaraldehyde for 4 hours in the refrigerator. He then rinses tissues in cold buffered sucrose following which they are post-fixed in Palade's osmium tetroxide for 1 hour. Tissues may be left in buffered sucrose solution in the refrigerator for longer periods if desired. The advantages of this method are improved penetration and preservation of sor ? enzyme activity.

Buffered glutaraldehyde (pH 7.3–7.4) Karlsson and Schultz (1965)

Sod. dihydrogen phosphate ($NaH_2PO_4.H_2O$)	3.31 g
Disodium hydrogen phosphate ($Na_2HPO_4.7H_2O$	33.77 g

Distilled water . . . . 925 ml
25 per cent Commercial glutaraldehyde . . . . 100 ml

Buffered sucrose solution (ph 7.3)

Sorensen's phosphate buffer pH 7.3–7.4 . . . . . 100 ml
Sucrose . . . . 4.5 g

The sucrose content is varied for different tissues, 4.5 per cent being a useful concentration for general use.

Since these fixatives penetrate very slowly into tissues, it is essential that the tissue blocks be cut into very small pieces, usually less than 1 mm in diameter. This must be done quickly, as soon as the tissues are removed, otherwise changes in all structures occur which although invisible in the light microscope, cause serious artifacts in the electron microscope. Clean razor blades are useful in dicing the tissue, care being taken to avoid compressing or otherwise distorting the tissues.

Fixation times depends on the size of the blocks and the type of tissue. Most tissues such as kidney and liver (in blocks less than 1 mm in diameter) will be well fixed in 1 hour. For special material such as brain, or botanical material, the reader should refer to papers by those working in these fields.

Dehydration

After fixation, the tissues are washed briefly in distilled water. Small glass vials or test tubes are suitable. Dehydration may be carried out in ascending concentrations of alcohol, starting with 30 per cent and progressing to 95 per cent in 3 or 4 steps of 5 minutes each. Acetone may also be used. While in 95 per cent alcohol the tissues should be examined under a dissecting microscope and further dissected if the blocks are too large. The tissues will be quite hard and easy to cut. They are transferred with Pasteur pipettes to 100 per cent alcohol and are now ready for embedding.

Embedding

In order to cut the very thin sections which will be transparent to the electron beam, it is necessary to embed the tissues in plastic. There are now two commonly used plastics for this purpose. It is recommended that one of these should be used routinely so that its characteristics become familiar, and faulty embedding can be recognized when it occurs. The methacrylates are easy to use, tissues are well infiltrated and the blocks are easy to section. The plastic will evaporate in the heat of the electron beam, however, so that the section collapses and resolution suffers even although contrast is good. The epoxy resins are viscous solutions and penetration of the tissues may be difficult. The blocks are more difficult to section but the plastic is stable and resolution is good. Contrast is poor, because these plastics have high electron scatter and do not evaporate in the electron beam. It is usually necessary to stain the sections.

Polyester Embedding (Sjostrand, 1967)

The polyester, Vestopal W, may be used as an embeddant, and is thought by many to give superior resolution. For use, the Vestopal W is mixed with an initiator (tertiary butyl perbenzoate) and an activator (cobalt naphthenate). Acetone should be used for dehydration instead of ethanol, because Vestopal W is insoluble in ethanol.

The Vestopal, initiator and activator, must be stored in a refrigerator and protected from exposure to light. The Vestopal can be stored for several months, but neither the initiator nor the activator are usually stable for more than 2 months; should the finished blocks be soft it is most probably due to inactivation of the latter.

Method

Dehydration

30 per cent Acetone	. 15–30 minutes.
50 per cent Acetone	. 15–30 minutes.
75 per cent Acetone	. 15–30 minutes.
90 per cent Acetone	. 30–60 minutes.
100 per cent Acetone (dried over $CuSO_4$) .	. 30–60 minutes.

Infiltration

1 part Vestopal W/3 parts acetone	. 30–60 minutes.
1 part Vestopal W/1 part acetone . . .	. 30–60 minutes.
3 parts Vestopal W/1 part acetone	. 30–60 minutes.
Vestopal A/+ 1 per cent initiator + 0.5 per cent activator	. 12–24 hours.

Embedding

Transfer tissues to gelatin capsules filled with Vestopal W to which has been added 1 per cent initiator (tertiary butyl perbenzoate) and 0.5 per cent activator (cobalt naphthenate) and leave at 60°C for 24–48 hours to polymerize and harden. When preparing the mixture the initiator is first well mixed in the Vestopal W before adding the activator and it is important that the mixing is performed carefully and thoroughly to avoid irregular polymerization: the mixture does not keep for more than a few hours at room temperature.

Kurtz (1961) does use ethanol to dehydrate tissue, by transferring it from absolute ethanol to styrene (3 changes over a period of 1 hour, then into a mixture of equal parts of styrene and Vestopal W for 30 minutes followed by pure Vestopal W for 4–48 hours).

Embedding in Epoxy Resins

Epon 812 (Shell Oil Co.) is most commonly used, having replaced Araldite which has high viscosity. The following method is after Luft.

Solution A
Epon 812 . . . . : . . . . 62 ml
DDSA (Dodecenyl succinic anhydride) 100ml

Solution B
Epon 812 . . . . 100 ml
MNA (Methyl nadic anhydride) . . . . : 89 ml

These solutions keep well in the refrigerator. The embedding mixture is made by adding together A and B in proportions depending on the degree of hardness desired. Seven parts of solution A to 3 parts of solution B will give a soft block, 3 parts of A to 7 of B will give a hard block. Intermediate mixtures may be used. It is best to experiment and find which is most suitable for the type of tissue and microtome being used. A suitable catalyst is added to the mixture, such as 1.5 per cent DMP–30*. Only sufficient plastic for immediate use should be prepared.

Infiltration with the fairly viscous plastic solution is best done through propylene oxide which does not interfere with polymerization. After dehydration in absolute alcohol the tissues are transferred to propylene oxide for 1 hour with frequent swirling. Most of the propylene oxide is then decanted leaving a few ml in the bottom of the vial with the tissue. An equal amount of the prepared Epon mixture is added with continuous swirling until the solutions have completely mixed. After half an hour the solution is decanted and replaced by the Epon mixture. This should be placed in the oven for an hour with frequent swirling. Suitable gelatin capsules are then filled with the plastic and the tissues are transferred with a Pasteur pipette to the top

*Rhom and Haas Co., Philadelphia, Penna., U.S.A.

of the capsules. They are returned to the oven during which time the tissues settle through the plastic to the bottom of the capsules. They may be centered with a probe. Polymerization may be obtained overnight at 50–60°C. The excess plastic may be trimmed away from the hardened blocks with razor blades or, since the plastic is quite tough, with a dental burr.

Sectioning

There are many models of microtomes to choose from which have been designed to cut the very thin sections needed for electron microscopy. The operation of these instruments requires a good deal of skill which can only be obtained by practice, so that there is little to be gained by describing here the techniques: the main requirement, apart from well embedded tissues, is patience.

Diamond knives may be used for methacrylate blocks, but are unsuitable for epoxy resins unless the knife angle is less than 45 degrees. Glass knives are cheap and easily made by the following method.

Plate glass, $^3/_{16}$ or $^1/_4$ inches thick, is scored into strips about 1½ inches wide. A break is started by tapping the reverse side of the score line, and the strip is broken off by pressure over an applicator stick. The broken surface is examined and only those strips which have a smooth surface are selected; have a large waste-basket nearby. The good strips are scored into rhombic pieces with an angle of 45 degrees or 50 degrees. The score line should not be carried closer than ¼ inch to the good edge of the strip. The knives can be broken off with glazier's pliers, or by touching the centre of the score line with a Pyrex glass rod heated to white heat in an oxygen flame. Each knife edge should be examined in a low power microscope and the parts of the edge which are free of nicks or fine lines noted. A suitable water trough is attached to the knife to catch the sections as they are cut. The sections are then picked up on previously prepared grids *(see below)*.

Glass Knife Makers

The introduction of glass-knife-making machines, such as the LKB which we use, has greatly facilitated the preparation of first-class glass knives with very little practice. The use of these machines, together with standardized glass strips, has done much to simplify the technique (and remove many of the frustrations) of ultra-thin sectioning.

Staining

The sections (supported by the grids) may be stained by a variety of heavy metal salts to increase the contrast. Most of these are non-specific, and indeed the nature of the combination with tissue constituents is not understood. One of the most useful is lead hydroxide. Lead hydroxide solutions may be prepared from sodium hydroxide and lead acetate, but are unstable and tend to cause heavy contamination due to deposits of lead carbonate. The method described by Karnovsky is recommended where the lead hydroxide is used in a highly alkaline solution which is easy to prepare and gives strong staining without contamination.

Lead Citrate Staining (Reynolds, 1963)

Lead citrate solution

Lead nitrate $(Pb(NO_3)_2)$ . . . . 1.33 g
Sodium citrate $(Na_3C_6H_5O_7.2H_2O)$ . . . . . . 1.76 g
Distilled water . . . . 30 ml

Place the above solution in a 50 ml volumetric flask, shake for 1 minute, then intermittently for 30 minutes. Add 8 ml of IN sodium hydroxide (NaOH), and make up to 50 ml with distilled water; mix by inversion. If the solution is turbid, centrifuge until clear immediately before use.

Staining the grids. – The grids are stained as described below for lead hydroxide. The stain is diluted from 1:5 to 1:1,000 with 0.01N NaOH and grids are stained for 4–5 minutes. After staining the grids are washed in 0.02N NaOH and distilled water.

Lead Hydroxide Stain (Karnovsky, 1961)

Add a spatula tip of lead oxide to 20 ml of 1 N NaOH and boil gently for 15 minutes. Cool and filter. Dilute the filtrate 50 or 100 times with distilled water. *The sections are stained conveniently on a paraffin surface* which is covered with a small Petri dish. A drop of the staining solution is placed on the paraffin and the grid floated on the drop with the section down in contact with the solution. Staining time is 5–30 minutes. The grid is then washed by transferring rapidly through three beakers of distilled water using jeweller's forceps, and then dried on filter paper.

Uranyl Acetate Staining

A 1 per cent solution is prepared by adding 1 g of uranyl acetate to 100 ml double distilled water, which is shaken well and allowed to sit for a day or two; the solution should be centrifuged before use. Staining is carried out as described above for 2–60 minutes (we use 5 minutes). Grids are rinsed well in distilled water.

Increased contrast can be obtained by double staining with lead citrate (above). The uranyl acetate is particularly good for fibrous proteins, it may also be used in alcoholic solution before the embedding process. Movat has described the use of silver proteinate (Protargol) and the silver methenamine stain which reacts with components of basement membranes. Many other heavy metals and organic dyes have been used in electron microscopy and many are useful in special circumstances. The original articles should be consulted.

Preparation of Electron Microscope Grids

The thin sections prepared for the electron microscope must be supported on plastic-covered grids to prevent their collapse in the heat of the electron beam. The grids are purchased and must be covered by a very thin film of plastic. Either formvar or collodion may be used and there are two principal methods of preparation.

Method I

Formvar is dissolved in ethylene dichloride at a concentration of 0.5 per cent, or collodion is dissolved in amyl acetate in a concentration of 1 per cent. Clean glass slides are placed for a few minutes in the plastic solution and then withdrawn and allowed to dry in a vertical position. A thin layer of plastic is formed over the surface of the glass slide as the solvent evaporates. The film, attached to the glass slide, is scored with a razor blade, into squares large enough to cover the electron microscope grids being used, and these are floated off on to the surface of a dish of distilled water, and then picked up one by one to cover one side of the grids. Each grid must be examined to ensure that it is completely covered by a smooth film.

Method II

A solution of 2 per cent collodion in amyl acetate is prepared. A large funnel is filled with water and a stainless steel screen is placed in the water near the top. As many as a hundred grids are placed on the screen. A drop of the collodion solution is allowed to fall on the surface of the water. This quickly spreads out and, as the solvent evaporates, a thin layer of collodion is left behind on the surface of the water. If the film is smooth with an even silvery colour it is probably suitable. The water is then drained out through the bottom of the funnel allowing the collodion film to settle over the screen and grids. After drying the grids are picked off and examined.

References

Batty, I., and Walker, P. D. (1964). *J. Path. Bact.*, **88**, 32.

Beck, C. (1938). *The Microscope.* London; Beck.

Berg, N. O. (1951). *Acta path. Scand., Suppl.,* 90.

Burch, C. R., and Stocks, J. P. P. (1942). *J. Sci. Instr.,* **19**, 71.

Burstone, M. S. (1960). *J. nat. Cancer Inst.,* **24**, 1199.

Chadwick, C. S., McEntegort, M. C., and Nairn, R. C. (1958). *Lancet,* **1**, 412.

Chick, E. W. (1961). *Arch. Derm. Syph.,* **83**, 305.

Coons, A. H. (1951). *Fed. Proc.,* **10**, 558.

– (1956). *Int. Rev. Cytol.,* **5**, 1.

– Creech, H. J., and Jones, R. N. (1941). *Proc. Soc. exp. Biol., N.Y.*

– and Kaplan, M. H. (1950). *J. exp. Med.,* **91**, 1.

– Leduc, E. H., and Connolly, J. M. (1955). *J. exp. Med.,* **91**, 1.

Creech, H. J., and Jones, R. N. (1940). *J. Amer. chem. Soc.,* **62**, 1970.

Culling, C. F. A. (1950). *Bull. Inst. med. Lab. Tech.,* **15**, 1, 8.

– (1961). *Arch. Path.,* **71**, 76.

– (1967). *Nature, Lond.,* **214**, 1140.

– (1974). *Handbook of Histopathological and Histochemical Techniques,* 3rd ed. London; Butterworths.

– and Saunders, (1960). *Can. med. Ass. J.,* **83**, 530.

– and Vassar, P. S. (1961). *Can. med. Ass. J.,* **85**, 142.

Fennell, R. H., Rodnam, G. P., and Vazquez, J. J. (1962). *Lab. Invest.,* **11**, 24.

Gersh, I. (1956). *J. Biophys. Biochem. Cytol.,* **2** (Suppl.), 37.

Goldstein, G., Slizys, I. S., and Chase, M. W. (1961). *J. exp. Med.,* **114**, 89.

Grunbaum, B. W., and Wellings, S. R. (1960). *J. Ultrastructure Res.,* **4**, 73.

Hellstrom, H. (1934). *H. Brit. Chem. Abst.,* 724.

Hicks, J. D., and Matthaei, E. J. (1958). *J. Path. Bact.,* **75**, 375.

REFERENCES

Higgins, G. C., and Jones, L. A. (1953). *Phot. Sci. Tech.,* **198,** 55.
Hiramoto, R., Engel, K., and Pressman, D. (1958). *Proc. Soc. exp. Biol. Med.,* **97,** 611.
Holt, S. J., and Hicks, R. Marion. (1961). *J. Biophys. Biochem. Cytol.,* **11,** 31 and 47.
Karlsson, U., and Schultz, R. (1965). *J. Ultrastruct. Res.,* **12,** 160.
Karnovsky, M. J. (1961). *J. Biophys. Biochem. Cytol.,* **11,** 729.
Kawamura, A. (1969). *Fluorescent Techniques.* Tokyo; University of Tokyo Press.
Kempson, D. A., Thomas, O. L., and Baker, J. R. (1948). *Quart. J. Micr. Sci.,* **89,** 3.
Kuper, S. W. A., and May, R. (1960). *J. Path. Bact.,* **79,** 59.
Mellors, R. C., Nowoslawski, A., and Korngold, L. (1961). *Amer. J. Path.,* **39,** 533.
Moor, H., and Mühlether, K. (1963). *J. cell. Biol.,* **17,** 3.
– and Branton, D. (1964). *J. Ultrastruct. Res.,* **11,** 401.
Nairn, R. C. (1962). *Fluorescent Protein Tracing.* Edinburgh; Livingstone.
Padawar, J. (1968). *J. R. micr. Soc.,* **88,** 305.
Popper, H. (1944). *Physiol. Rev.,* **24,** 205.
Pressman, D., Yagi, V., and Hiramoto, R. (1958). *Int. Arch. Allergy, N.Y.,* **12,** 125.
Reynolds, E. S. (1963). *J. cell Biol.,* **17,** 208.
Riggs, J. L., Seiwald, R. J., Burckhelter, J., Downs, C. H., and Metcalf, T. G. (1958). *Amer. J. Path.,* **34,** 1081.
Sabatini, D. D., Bensch, K., and Barnnett, R. J. (1963). *J. cell. Biol.,* **17,** 19.
Singer, S. J., and Schick, A. F. (1961). *J. Biophys. Biochem. Cytol.,* **90,** 519.
Sjöstrand, F. S. (1967). *Electron Microscopy of Cells and Tissue.* New York and London; Academic Press.
Taylor, H. E., and Shepherd, W. E. (1960). *Lab. Invest.,* **9,** 603.
Vassar, P. S., and Culling, C. F. A. (1959). *Arch. Path. (Chicago),* **67,** 120.
– – (1961). *Amer. J. clin. Path. (Chicago),* **36,** 244.
– – (1962). *Arch. Path. (Chicago),* **73,** 59.
– and Saunders, A. M. (1960). *Arch. Path. (Chicago),* **69,** 613.
Vazquez, J. J., and Dixon, F. J. (1957). *Lab. Invest.,* **6,** 205.
Weller, T. H., and Coons, A. H. (1954). *Proc. Soc. exp. Biol., N.Y.,* **86,** 789.
Wellman, K. F., and Teng, K. P. (1962). *Can. med. Ass. J.,* **87,** 137.
White, R. G. (1957). *Proc. R. Soc. Med.,* **50,** 953.
Wignall, N., Culling, C. F. A., and Vassar. P. S. (1961). *J. clin. Path.,* **36,** 469.

Index

143

INDEX